Don't Throw in the Trowel!

Vegetable Gardening Month by Month

Good Quality S
5 in., 10c. each; $1.00 per doz.
7 in., 15c. each;

Easy-Growing Gardening Series, Vol. 1
by
Rosefiend Cordell

Rosefiend Publishing
Copyright © 2016 Melinda R. "Rosefiend" Cordell

All rights reserved. Although the author has made every effort to ensure that the information in this book was correct at press time, the author does not assume and hereby disclaims any liability to any party for any loss, damage, or disruption caused by errors or omissions, whether such errors or omissions result from negligence, accident, explosions, drunken raccoons, the wrath of Khan, tiny bugs, bulldozer races, or any other cause.

**Sign up for my newsletter and get a free book!
Visit me at melindacordell.com**

Edelweiss.

DEDICATION

To Grandma Annamarie and Grandma Mary

I used to call you guys on the phone and say, "How do you guys do this??" and you'd always tell me how vegetable gardening worked, because that's what grandmas do. I love you both.

Contents

DEDICATION	3
INTRODUCTION	7
HORTICULTURING IN THE VEGETABLE GARDEN	7
JANUARY	12
Save Time and Trouble With Garden Journals	12
Planning the Vegetable Garden	19
A Vegemaniac in Winter	21
Checking Seed Germination Rates	26
Laying Out the Garden the Square Foot Way	28
January List o' Things to Do!	31
FEBRUARY	34
How to Start Your Seeds	34
Cold Frame Setup	40
Spring Supply List	41
February List o' Things to Do!	46
MARCH	48
Turn Your Weeds into Green Manure	49
Me Versus the Tiller	51
Other Tilling Hints and Helps	54
Seeding, Planting, Transplanting Outside	57
Ways to Discourage Rabbits and Squirrels	58
WHEN DO I PLANT AND SEED IN MARCH?!	61
March List o' Things to Do!	64
APRIL	67
Planting Potatoes	67
It's a Bug-Eat-Bug World	71
Herbs in the Garden	75
Early April Planting Guide	78
April Showers Means May Flowers … and Weeds!	80
April Planting Tips	81
April List o' Things to Do!	82
MAY	88
A Quick Planting Guide to Corn	88

Tips on Growing Melons	93
Vertical Gardening: Put Your Crops in the Air	95
Mexican Bean Beetles	96
Mulch Down The Weeds	98
Stop the Grasshopper Invasion	100
Spotted Cucumber Beetles are Evil	104
May List o' Things to Do!	107
JUNE	112
Don't Become a Statistic: Avoid Heat Exhaustion	112
Potato Blight Also Affects Tomatoes	116
Kill the Right Bug	121
Squash Bugs	123
Fusarium, Bacterial, and Other Vegetable Wilts	126
June List o' Things to Do!	130
JULY	133
Sometimes Plants Need a Break from the Heat	133
When Weeds Get Out of Control	135
Practical Harvest Tips	137
Bitter Melons and Other Asian Squashes	139
July List o' Things to Do!	148
AUGUST	152
Late-Summer Planting for Fall Vegetables	152
Fix Weeds with Newspaper Mulch	155
August List o' Things to Do!	158
SEPTEMBER	160
Getting Tomatoes Ready for Harvest	161
Vegetables in Winter	163
Drying Fruits and Vegetables	169
September List o' Things to Do!	177
OCTOBER	181
Pumpkin Season!	181
Amendments To Build Rich Garden Soil	186
October List o' Things to Do!	189
NOVEMBER	192
Collecting Seeds For Next Year	192
Taking Soil Samples	195
Heirloom Vegetables – A Long and Colorful Tradition	198

November List o' Things to Do!	201
DECEMBER	204
Composting: Learn the Basics	204
Tool Care Tips	209
December List o' Things to Do!	211
BOOK RECOMMENDATIONS	214
Tomato Varieties	217
Determinate vs. indeterminate	220
Starting Tomatoes from Seed	220
CREATING A FOUR-SEASON PERENNIAL GARDEN	226
Rules of (Green) Thumb for Garden Design	232
ABOUT THE AUTHOR	236

INTRODUCTION

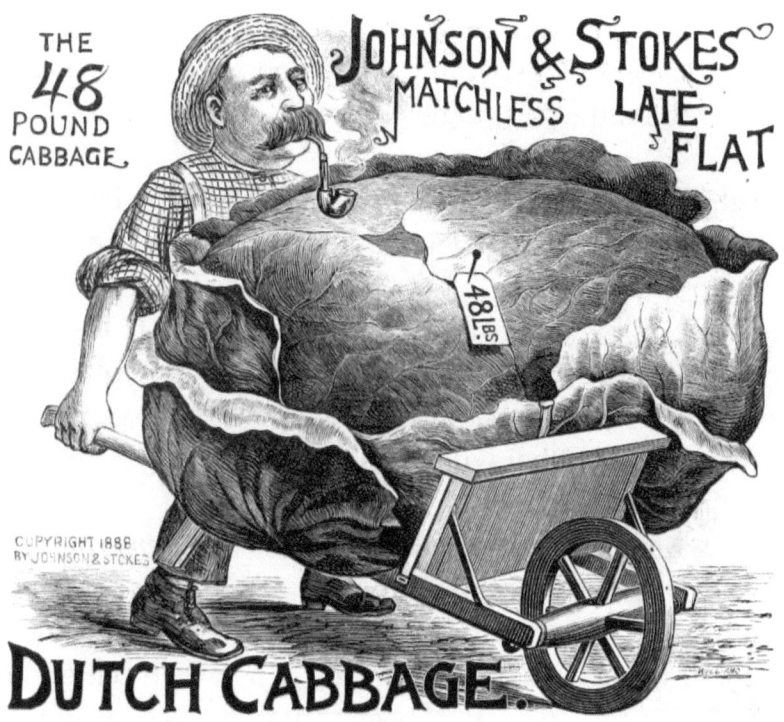

HORTICULTURING IN THE VEGETABLE GARDEN
Or, Gardeners learn by trowel and error.

Two things:

First: You know more about gardening than you think.

Second: A garden – the soil – plants – all of these are very forgiving. When it comes down to it, you can do pretty much anything to these. (Though, actually, bulldozer races through the garden are out of the question.)

A garden is forgiving. Plants are built to put up with a lot of nonsense. They obviously can't get up and walk away, so they're made to endure. It's part of their nature.

In a vegetable garden, as with anything in life, the best thing to do is to focus on raising vegetables that you really like, stuff that you can't wait to eat when harvest season rolls around. Concentrate on the things in the vegetable garden that you really enjoy. That, more than anything, is the secret to gardening success. If you're crazy about something, it will show. If you don't care about it – that will also show!

I used to write a gardening column for the *St. Joseph (Mo.) News-Press*. I was a natural choice for the job – I had worked in horticulture for most of my life, starting when I was a senior in high school when I got a part-time job at a garden center. I got a degree in horticulture and began working as a municipal horticulturist, where I had to learn the ropes quickly. As city horticulturist, I took care of 20-28 gardens all around the city, and did the heavy labor pretty much on my own, including a lot of the mulching and digging and weeding and shearing and deadheading and fertilizing. I also had 300 roses in the rose garden, and hundreds of trees. Spring was nuts. So was the rest of the year.

But every other week, I would share my experiences with a small bunch of stalwart readers, who were just as kind as they could be. It always amazed me when somebody would recognize me from my picture in the paper. I always figured my grandmas read my columns and that was it!

I'd been looking at the old columns for years, wondering about putting them out into the world again. I am finally biting the bullet. This book is put together from my old columns, though I've added a lot of new stuff and have updated information. I will be writing a series of books using these columns – Vegetable Gardening, Roses, Annuals and Perennials, Houseplants, Trees and Shrubs, and Soilbuilding. It should be great fun.

Don't Throw in the Trowel! A Month-by-Month Guide to Vegetable Gardening is aimed specifically for Midwest gardeners in Zone 5, though these tips will apply to those in Zones 4 and 6. Gardeners in Zone 4 will be about a week or two behind, while gardeners in Zone 6 will be about a week or two ahead. For best results, double-check your planting and growing dates with your local University Extension center.

This book will be a month-by-month guide to vegetable gardening, and is organized as such. The prep is just as important as the planting, and in some ways more important, because a good plan, a garden notebook, and a little off-season work will save you a lot of trouble down the road.

This book is also aimed those who know their way around a garden, and beginners. I'll cover simple topics and more complex ones, and a bunch of stuff in between. I'll tell you one thing: It doesn't matter whether you're just starting out (Protip: Plant it with the green side up) or if you've been horticulturing all your life – there is always, always, always something new to learn. That's the beauty and the wonder of the world, that there's always new knowledge – or old knowledge brought to light – to discover.

Also, there are few hard rules in gardening. You are dealing with living things, and as with any other living beings, they have a mind of their own. The only difference is that plants can't run and they can't talk back. But they show it when they're content and burying your garden in tomatoes, and they'll show it when they're having a rough time and catching diseases or bug infestations or looking stunted and wilty. Then you get to figure out why the plant is stressed and find ways to correct the problem, to help your plant regain its health and fight off diseases and bugs.

I use organic methods in this book, because they are cheap, fairly easy, and they work very well. Organic gardening is all about improving the health of the soil. When you have a healthy soil, then you have healthy plants. A healthy soil is filled with organic matter, which you can get from compost and from a thick mulch of chopped leaves, grass clippings, old hay or straw, wood chips, or whatever is handy and shredded. These materials, as they decay, add nutrients to the soil. These materials also create a soil busy with microorganisms, tiny insects and arthropods, fungi, and earthworms that mix the soil and incorporate the organic matter in more deeply. The work of all of these soil organisms changes organic matter into a form that the plant roots can easily absorb.

Artificial fertilizer is like giving your plant a multivitamin. Organic material is like feeding your plant a healthy diet at all times.

I don't object to using chemicals occasionally. When I was horticulturing for the Parks Department, I was almost always working as a staff of one, so I used quick fixes – I used systemic fungicide in the rose garden, and I used Round-Up to knock down weeds quickly. I used only as much as I needed – nothing extra – and was careful. With the Round-Up, I sprayed only a little bit directly on the plant instead of waving my sprayer wand hither and yon. In among the flowers, I developed a knack of wiping the applicator, which had a few drops of herbicide on it, directly on the weed leaves. This usually did the trick. Use the smallest possible amount – this saves money and keeps extra chemicals out of the soil and groundwater.

The great thing about gardening is that you can get everything wrong and still come out all right, mostly. Plants are forgiving (except when they croak, then they come back with murderous vendettas – okay, maybe they don't). Soil is easygoing, as I hope this guide will be. If something doesn't work, pull it up and try again, or use a different method next year. The important thing is that we keep trying – and keep having fun with it.

JANUARY

Save Time and Trouble With Garden Journals

When I worked as a municipal horticulturist, I took care of twelve high-maintenance gardens, and a number of smaller ones, over I-don't-know-how-many square miles of city, plus several hundred small trees, an insane number of shrubs, a greenhouse, and whatever else the bosses threw at me. I had to find a way to stay organized besides waking up at 3 a.m. to make extensive lists. My solution: keep a garden journal.

Vegetable gardeners with an organized journal can take control of production and yields. Whether you have a large garden or a small organic farm, it certainly helps to keep track of everything in order to beat the pests, make the most of your harvest, and keep up with spraying and fertilizing.

Keeping a garden journal reduces stress because your overtaxed brain won't have to carry around all those lists. It saves time by keeping you focused. Writing sharpens the mind, helps it to retain more information, and opens your eyes to the world around you.

My journal is a small five-section notebook, college ruled, and I leave it open to the page I'm working on at the time. The only drawback with a spiral notebook is that after a season or two I have to thumb through a lot of pages to find an earlier comment. A small three-ring binder with five separators would do the trick, too. If you wish, you can take out pages at the end of each season and file them in a master notebook.

I keep two notebooks – one for ornamentals and one for vegetables. However, you might prefer to pile everything into one notebook. Do what feels comfortable to you.

These are the five sections I divide my notebooks into – though you might use different classifications, or put them in different orders. Don't sweat it; this ain't brain surgery. Feel free to experiment. You'll eventually settle into the form that suits you best.

First section: To-do lists.
This is pretty self-explanatory: you write a list, you cross off almost everything on it, you make a new list.

When I worked as horticulturist, I did these lists monthly. I'd visit all the gardens I took care of. After looking at anything left unfinished on the previous month's list, and looking at the garden to see what else needed to be done, I made a new, comprehensive list.

Use one page of the to-do section for reminders of things you need to do next season. If it's summer, and you think of some chores you'll need to do this fall, make a FALL page and write them down. Doing this has saved me lots of headaches.

Second section: Reference lists.
These are lists that you'll refer back to on occasion.

For example, I'd keep a list of all the yews in the parks system that needed trimmed, a list of all gardens that needed weekly watering, a list of all places that needed sprayed for bagworms, a list of all the roses that needed to be babied, etc.

I would also keep my running lists in this section, too – lists I keep adding to. For instance, I kept a list of when different vegetables were ready for harvest – even vegetables I didn't grow, as my friends and relatives reported to me. Then when I made a plan for my veggie garden, I would look at the list to get an idea of when these plants finished up, and then I could figure out when I could take them out and put in a new crop. I also had a list of "seed-to-harvest" times, so I could give each crop enough time to make the harvest date before frost.

You can also keep a wish list – plants and vegetables you'd like to have in your garden.

Third section: Tracking progress.

This is a weekly (or, "whenever it occurs to me to write about it") section as well.

If you plant seeds in a greenhouse, keep track of what seeds you order, when you plant them, when they germinate, how many plants you transplant (and how many survive to maturity), and so forth.

When you finish up in the greenhouse, use these pages to look back and record your thoughts – "I will never again try to start vinca from seeds! Never!! Never!!!" Then you don't annoy yourself by forgetting and buying vinca seeds next year.

You can do the same thing when you move on to the vegetable garden – what dates you tilled the ground, planted the seeds, when they germinated, and so forth. Make notes on yields and how everything tasted. "The yellow crooknecks were definitely not what I'd hoped for. Try yellow zucchini next year."

Be sure to write a vegetable garden overview at season's end, too. "Next year, for goodness' sake, get some 8-foot poles for the beans! Also, drive the poles deeper into the ground so they don't fall over during thunderstorms."

During the winter, you can look back on this section and see ways you can improve your yields and harvest ("The dehydrator worked great on the apples!"), and you can see which of your experiments worked.

Fourth section: Details of the natural world.

When keeping a journal, don't limit yourself to what's going on in your garden. Track events in the natural world, too. Write down when the poplars start shedding cotton or when the Queen's Anne Lace blooms.

You've heard old gardening maxims such as "plant corn when oak leaves are the size of a squirrel's ear," or "prune roses when the forsythia blooms." If the spring has been especially cold and everything's behind, you can rely on nature's cues instead of a calendar when planting or preventing disease outbreaks.

Also, by setting down specific events, you can look at the journal later and say, "Oh, I can expect little caterpillars to attack the indigo plant when the Johnson's Blue geranium is blooming." Then next year, when you notice the buds on your geraniums, you can seek out the caterpillar eggs and squish them before they hatch. An ounce of prevention, see?

When I read back over this section of the journal, patterns start to emerge. I noticed that Stargazer lilies bloom just as the major heat begins. This is no mere coincidence: It's happened for the last three years! So now when I see the large buds, I give the air conditioner a quick checkup.

Fifth section: Notes and comments.
This is more like the journal that most people think of as being a journal – here, you just talk about the garden, mull over how things are looking, or grouse about those supposedly blight-resistant tomatoes that decided to be contrary and keel over from blight.

I generally put a date on each entry, then ramble on about any old thing. You can write a description of the garden at sunset, sketch your peppers, or keep track of the habits of bugs you see crawling around in the plants. This ain't art, this is just fun stuff (which, in the end, yields great dividends).

Maybe you've been to a garden talk on the habits of Asian melons and you need a place to put your notes. Put them here!

This is a good place to put garden plans, too. Years later I run into them again, see old mistakes I've made, and remember neat ideas I haven't tried yet.

Get a calendar.
Then, when December comes, get next year's calendar and the gardening journal and sit down at the kitchen table. Using last year's notes, mark on the calendar events to watch out for -- when the tomatoes first ripen, when the summer heat starts to break, and when you expect certain insects to attack. In the upcoming year, you just look at the calendar and say, "Well, the squash bugs will be hatching soon," so you put on your garden gloves and start smashing the little rafts of red eggs on the plants.

A garden journal can be a fount of information, a source of memories, and most of all, a way to keep organized. Who thought a little spiral notebook could do so much?

A Victory Garden for a Family of Five
On a Plot 25 x 50 ft.

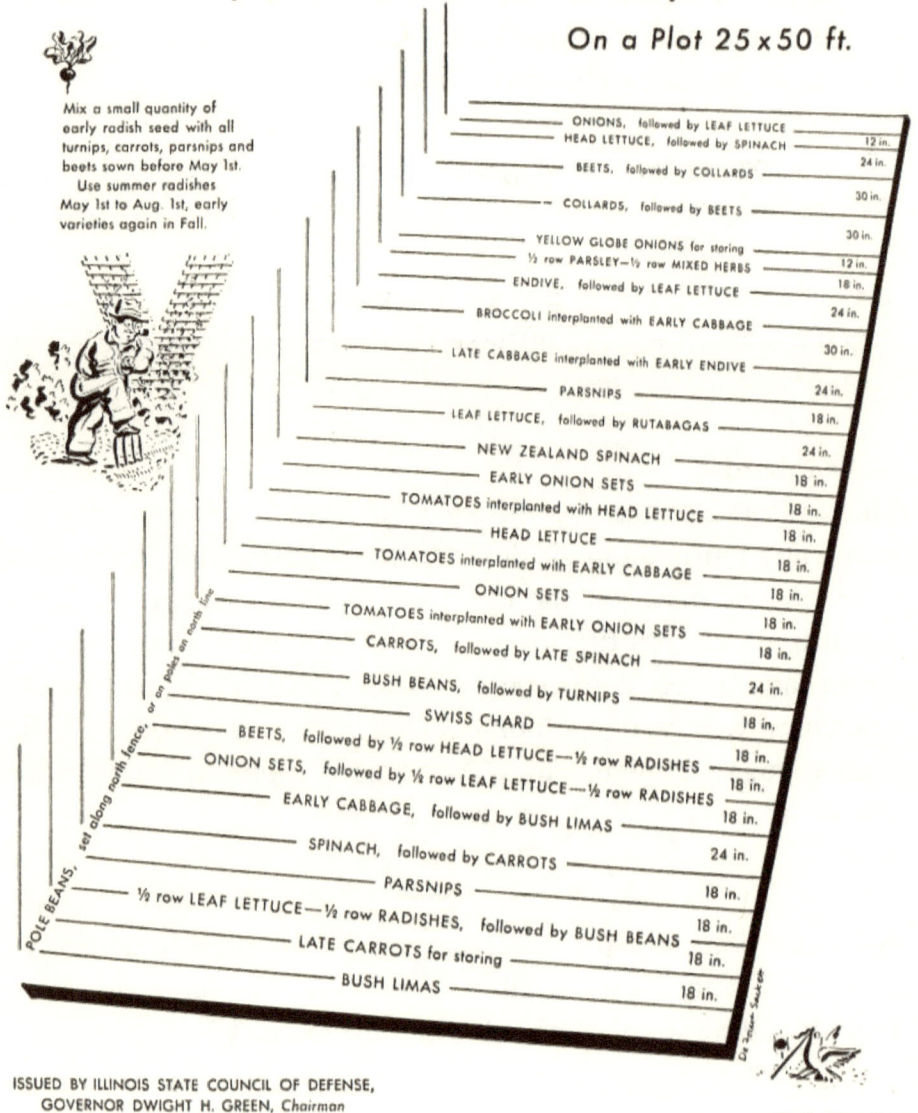

Mix a small quantity of early radish seed with all turnips, carrots, parsnips and beets sown before May 1st. Use summer radishes May 1st to Aug. 1st, early varieties again in Fall.

Row	Width
ONIONS, followed by LEAF LETTUCE	12 in.
HEAD LETTUCE, followed by SPINACH	24 in.
BEETS, followed by COLLARDS	30 in.
COLLARDS, followed by BEETS	30 in.
YELLOW GLOBE ONIONS for storing	12 in.
½ row PARSLEY—½ row MIXED HERBS	18 in.
ENDIVE, followed by LEAF LETTUCE	24 in.
BROCCOLI interplanted with EARLY CABBAGE	30 in.
LATE CABBAGE interplanted with EARLY ENDIVE	24 in.
PARSNIPS	18 in.
LEAF LETTUCE, followed by RUTABAGAS	24 in.
NEW ZEALAND SPINACH	18 in.
EARLY ONION SETS	18 in.
TOMATOES interplanted with HEAD LETTUCE	18 in.
HEAD LETTUCE	18 in.
TOMATOES interplanted with EARLY CABBAGE	18 in.
ONION SETS	18 in.
TOMATOES interplanted with EARLY ONION SETS	18 in.
CARROTS, followed by LATE SPINACH	24 in.
BUSH BEANS, followed by TURNIPS	18 in.
SWISS CHARD	18 in.
BEETS, followed by ½ row HEAD LETTUCE—½ row RADISHES	18 in.
ONION SETS, followed by ½ row LEAF LETTUCE—½ row RADISHES	18 in.
EARLY CABBAGE, followed by BUSH LIMAS	24 in.
SPINACH, followed by CARROTS	18 in.
PARSNIPS	18 in.
½ row LEAF LETTUCE—½ row RADISHES, followed by BUSH BEANS	18 in.
LATE CARROTS for storing	18 in.
BUSH LIMAS	

POLE BEANS, set along north fence, or on poles on north line

ISSUED BY ILLINOIS STATE COUNCIL OF DEFENSE,
GOVERNOR DWIGHT H. GREEN, Chairman

Planning the Vegetable Garden

It's excitement time when you grab your stacks of seed catalogs and your pencils and your big sheets of paper (or, for those of you who are savvy enough to be reading this book on your tablet with a landscaping app). Planning the vegetable garden is when you take all those bright ideas dancing in your head and try to put them on paper (or on a screen).

Here are a few questions to ask that will help you through the process.

What vegetables does your family like? Check with the rest of the crew to see if there are any vegetables they'd like to see on the table. Sometimes they might surprise you. So you keep planting pumpkins and making nutritious pumpkin spice everything. Have you asked the rest of your family if they want pumpkin spice everything? If they say No, then sigh dramatically but plant something else. The old saying, "everything in moderation," certainly applies here.

What do you want to get out of the garden? Do you plan to do a lot of canning/dehydrating? Are you trying to lower your food budget? Make holiday decorations? Plan (and plant) accordingly.

Always start small. A garden can end up being a daunting task. You can always expand later. Or, if there's a new vegetable you want to try out, plant just a little bit of it instead of two long rows.

How are you going to cultivate the rows? Tiller, hoe, tractor? Put enough space between the rows to allow your implement of choice to fit. An additional note: if you had a lot of trouble keeping up with weeds last year, maybe this year you might make the rows just wide enough to admit the lawnmower. No, I'm serious, that lawnmower can be a lifesaver when your garden gets out of control.

Other tips:
Put all the perennial crops (asparagus, rhubarb, strawberries) together at one end to make everything easier to manage.

You can plan for succession crop, or a fall garden, and overlap these areas. For example, after you've harvested all the radishes, plant the area with tomatoes. After the spinach is done, plant cucumbers. That way, you can keep production going after each crop is finished.

In hilly areas, plant along the contours of the hill. Up and down planting will lead to erosion.

Plant rows running north and south. Otherwise, the plants will shade each other during the day.

Later on, make notes about what worked and what didn't, and put them in your gardening notebook for next year.

A Vegemaniac in Winter

January would be a really lousy month if we didn't have seed catalogs. There's so much ice on my garden that I can walk over the snow without breaking through the slick crust on top. So every day I skate to the mailbox, hoping for some good reading material to chase away these winter blues.

I love the romance inherent in growing heirlooms, and my seed catalogs are all from heirloom vegetable companies. In the catalogs, you find vegetables brought over from the old country by somebody's grandparents, varieties popular in the Middle East, vegetables bred by Trappist monks, and my favorite, garbanzo beans grown in the Fertile Crescent over 6,000 years ago. Whew!

My Seed Savers Exchange catalog (http://www.seedsavers.org/) has shown up already. I am so in love with their pictures of heirloom vegetables. I wish I could plant all of them, except I have only so much room in the garden. The "Tigger" melons are cute, all red-orange zigzags on a little yellow globe, and they are fragrant. "Queen Anne's Pocket Melon" has small, orange fruits with a powerful fragrance, though the white flesh is bland. The catalog says that women used to carry these in their pockets for the perfume. And maybe for a little snack, I suppose. But I'm probably going to get the yellow Asian melon, "Sakata's Sweet." These are baseball-sized, yellow melons with sweet flesh, and it can be grown on a trellis. (I need a small melon because I'm the only person in my household who would eat it.) They also have a melon called Collective Farm Woman. It is "an old Ukrainian variety, very popular on the island of Krim in the Black Sea." It looks pretty good. Decisions, decisions ….

I'm no fan of peppers – I use them in soup and that's it. Also, I'm not too good with hot food. I chewed up a jalapeño pepper once, and my whole face started watering and hot lava come out of my ears. I'm a wimp, sue me. But here's a "Fish" pepper that is about medium heat (which means it would make me spit flames), but my, is this ever a cute pepper. It's an African-American heirloom with variegated leaves. The peppers ripen in the most amazing blends of green, red, orange, and yellow. I'd be happy to grow it for looks alone.

I'm also interested in these small, colorful peppers with a heat rating of zero, but I'm having a hard time choosing just one. "Tequila Sunrise" has orange, carrot-shaped fruits. The miniature bell peppers (chocolate, red, and yellow) are cute. "Healthy" has carrot-shaped red fruits. I'll probably have to flip a coin with these.

I'd like to get fingerling potatoes, I think. Fingerlings are small, skinny potatoes (about the size of a finger) that are a gourmet's delight. Problem is, they also have gourmet prices. Most of them cost $12.50 for a 2½ pound bag. I like the looks of the Russian Banana, which is supposed to be a great potato. It's very resistant to scab, but unfortunately it has low tolerance against potato blight. The Rose Finn Apple (what a cool name) is more resistant. Then I look at the regular potatoes, and there's Caribe, which is a purple-skinned potato with white flesh. It's a super all-purpose potato with heavy yields, and good when you harvest it as a new potato. So many potatoes, so little time. Flip a coin again.

Now, if only every seed catalog could include, along with the nice pictures, actual taste samples of the vegetables listed. "Here, try some Early Silverline melons. Have a bite of an Oxheart carrot."

In some places, it's possible! If you'd like a taste sample of your potatoes before you buy them, Wood Prairie Farm in Bridgewater, Maine, offers a Potato of the Month Club. They send you three potato varieties every month in a ten-pound gift box, from September to April. You can also order fresh organic vegetables – roots like shallots, "Chantenay" carrots, parsnips, and so forth – through the winter. Wood Prairie Farm and Seed Savers Exchange also offer packages of dried beans. So you *can* "taste and try before you buy."

Before you place your order for garden seeds, take a look at the Garden Watchdog website. Garden Watchdog provides information on thousands of gardening vendors. Folks write in their comments on seed companies they've done business with, noting the quality of the plants and seeds sent, and how they fared with customer service.

The Garden Watchdog site also lists their top 20 gardening vendors – the ones considered the best of the best, with great products and super customer service. The list includes the Antique Rose Emporium down in Texas, Johnny's, Pinetree, and Baker Creek Heirloom Seed Co., which operates out of Mansfield, Mo.

However, time is running out for sending off for a seed catalog and getting an order in. (The spring rush for gardening vendors starts in January and ramps up into high gear in March, so get your orders in early and beat the rush!)

I can see why the movement toward heirloom varieties is gaining momentum. These guys are traveling the world to bring you stuff you'd never ordinarily see – vegetables sold in markets in Israel, or grown in Uzbekistan, or eaten in Taiwan. You can save the seeds and keep growing these same veggies year after year. They definitely beat what you'd buy at the grocery store! And I'm always interested in trying something unique.

When you start, start easy, mixing your regular crops with a few new varieties. Once you've soaked up what you've learned along the way, go crazy. As the song says, you're never going to survive unless you get a little crazy.

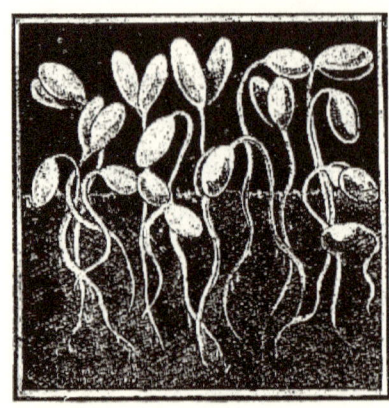

Checking Seed Germination Rates

Do you have some old seeds, and you're not sure if they're good or not? You can easily check their germination rate.

Take 10 seeds out of the packet as your sample. If you have a lot of seeds, 20 seeds will give you a more accurate sample. Get a damp paper towel – just moist, so it doesn't promote bacterial or fungal growth – and lay the seeds in a row on it.

Then fold the paper towel over the seeds, put it in a plastic bag and write today's date on it. (Leave the plastic bag open a tiny bit for air.) Then set it in a warm area for two to five days. Some seeds might take up to ten days to germinate. At any rate, once they all germinate, check them for fungal damage, and see how vigorous the new shoots are.

If the seed germination rate is less than 60 percent – that is, if fewer than 6 out of ten seeds germinate – then you might be better off getting a new packet of seeds. (Though you certainly could take them out and plant them with the new seeds – no point in wasting them!)

Laying Out the Garden the Square Foot Way

I was having a heck of a time planning my vegetable garden. I didn't need a lot of plants in the vegetable garden. Two yellow squash plants are more than enough for me, and five tomato plants are plenty (I'm the only one eating the tomatoes in this household). Of course when you're looking at seed catalogs, you want to buy everything between those two covers, but when you have a small garden, you have to pick and choose so you have room in your garden for everything you want to plant.

So I picked up a copy of *Square Foot Gardening* by Mel Bartholomew on the internet. What do you know, when I read the book, I found the system I'd been looking for! It's simple, and it's not wasteful.

First, the simple part. The backbone of the garden is the one-foot square. You take four of these one-foot squares and put them together in a grid. Separating these squares are paths, about a foot or two wide, so you don't walk on the soil inside the four-foot areas. (Soil compaction is bad news in the garden. I lay board walkways in the garden so I can keep from mashing the soil. A light, airy soil is very good for plants.)

When you draw the garden plan, you'll be drawing a lot of "windows," like the windows a little kid would draw on a picture of a house. In each "pane" you're growing one crop.

Spacing is pretty straightforward. In each one-foot square you can plant one, four, nine, or 16 plants, depending on their normal space requirements. So, you could grow one tomato, pepper, eggplant, or potato plant per square. Chard, bush beans, herbs, and large lettuce could be planted four per square. Smaller lettuces could be planted nine per square. Root crops like carrots, radishes, and beets could be planted 16 per square. Zucchini gets its own special category – it will take three feet of the four-foot square, and in the remaining area you could plant root vegetables or lettuce, i.e., plants that you'll have already harvested by the time the greedy zucchini sprawls all over.

Spacing isn't set in stone, of course. If you want to grow a monster tomato plant that feeds the world (wouldn't that be nice?), then set it in the middle of a four-foot square and get out of the way.

Plants like cantaloupe, pole beans, peas, and cucumbers can be grown on a trellis, which saves space, and also saves you from stooping over and hunting for the vegetables.

Next, the frugal part. You plant only as many seeds as you need. Instead of dumping an entire pack of lettuce seeds in the ground, you plant nine, or 16. And as for thinning: isn't it silly to grow all those seedlings, then pull up half of them? Or, on those occasions when you don't get around to thinning the seedlings, then end up with plants crowded together like sardines. They never reach their full size, so your crop is nowhere as good as it might have been.

Then, once you've written out a plan, you can grab those seed catalogs and order seeds to fit those spaces.

The benefits to square-foot gardening are many. To plan, you draw a bunch of windowpanes until you find an arrangement you like. Then scale those windowpanes to size and slap them on your garden plan. You plant only what you need. No waste! You don't walk all over the garden, just on the paths, and you can reach into the four-foot area and harvest without compacting the soil. If you put scrap lumber on the path to walk on, you can go into your garden even on rainy days. When the plants come up, put a thick layer of newspapers, or a heavy mulch, down around them, to keep weeds down. There's less work all around – and more time to enjoy gardening.

If you'd like more information, check out Mel's book, or look at his website.

January List o' Things to Do!

* **Check your tools to be sure everything is sharpened and oiled.** You might be surprised by rust on some tools if they were improperly put away last fall. (I'm speaking more for myself here, actually.) It's better to be surprised now than in early spring. Rub the wooden handles of your tools with paraffin. Check the tire on your wheelbarrow to see if it needs aired or replaced. You might even give the wheelbarrow a fresh coat of paint if it's starting to get rust spots.

* **On some warm day,** spray horticultural oil on your fruit trees, raspberry canes, and roses (those parts of the roses that aren't covered up). Give these plants a dose of lime sulfur, at winter strength, once a month. Studies have shown that roses sprayed

with lime sulfur through the winter have less disease through the year.

* **Add more mulch to your plants** if necessary. Do you have leaf piles from last fall? Run the lawn mower over the leaves to chop them and use them as mulch. I got ten big bags of leaves from my grandma, and about seven bags from my neighbors. I spread them all over the garden, then mowed and mowed. It took about an hour to mow them down properly, but the big heaps of leaves became a good layer of mulch.

Now I wish I had more leaves. The ground under the leaf mulch doesn't freeze brick-solid like unmulched ground. The earthworms have been busy under the leaves, coming up to the surface and leaving nutrient-rich castings everywhere – great natural fertilizer that takes no effort on your part. Okay, except for the mulching part. But hey, what good results you get!

* **Water your outside plants,** which lose moisture through their stems and their buds. Of course, do this when the temps are above freezing!

* **Wash your pots and trays** to get ready for seeding plants next month. Use hot, soapy water, then rinse them with a mild bleach solution and let them dry.

* **Organize your seeds.** See what you have available for this year, and make a note of it so you don't end up ordering a batch of seeds you already have.

* **During mild spells,** check your perennial vegetable and berry plants to see if any of them have "heaved" out of the ground, and gently press them back into the ground if they have.

FEBRUARY

How to Start Your Seeds

I got my seeds a month ago. They aren't doing much in the bag, and it's a little early to plant, but I just like to take them out and look at them. My daughter takes the bean seeds and goes around the house shaking them like maracas. I like to imagine what the garden might yield, though the little pessimist that lives in my mind keeps bringing to my daydreams images of blight and drought and cats leaving smelly presents in the mulch. And yet hope spring eternal.

Before you start your seeds, before you haul out the pots and the potting soil, take a moment to invest in a notebook. This notebook will be your best buddy through the whole process, recording things like seed-sowing dates, how well each batch of seeds germinated, seeding methods that worked (or didn't), and things you plan to do differently next year. If you already have a gardening notebook, use that. Notebooks are so handy.

Then, get out a calendar and look at the seed packet or the catalog and find out how long it takes from the time you plant the seed until the day you can put the plant in the ground. Then look at the calendar and count backward from your last frost date. That will be the date when you should plant those particular seeds, more or less.

Seed germination rates and seedling growth will vary from greenhouse to greenhouse. Some greenhouses have more light or warmth, while others don't. This will affect how soon your plants will be ready to go in spring. Sometimes the plants aren't ready to plant in May simply because you didn't get the seeds in early enough. By keeping track of your planting dates, you can readjust the "seed to finished product" time next year.

To start seedling flats, I use a sterile, seed-starting soilless mix that's high in vermiculite – seeding mixes tend to be a light and fluffy compared to regular potting soils – and get a bunch of trays (or flats). I prefer to use those black trays you see in greenhouses, the ones holding the six-packs or four-packs of flowers – but they are pricey! Clay pots for starting seeds require more soil than necessary, especially if you're going to end up transplanting the young plants into a different pot. Egg cartons are cheap and easy. Peat tablets (the ones that expand when you add water to them) are neat. If any of these methods work for you, then go for it. The rules you should follow are the rules that work best for you.

In my opinion, trays are the best for an operation where you need lots of plants and you need to work fast with them. Also, trays are easy to wash and store.

So, fill the tray about half-full of soilless mix and tamp it down to make a flat surface for the seeds. I used a small board, cut to fit the inside of the tray, to tamp the soil. If you don't have a board, your hands work fine.

Sowing seeds both large and small

Scatter large seeds across the top and cover them with soil. The smallest seeds can be the cheapest, but they can also be tricky to handle. Alyssum, which has small, flat seeds, will blow away if you sneeze.

When I worked at the greenhouse with George Ferbert, he showed me a little trick when seeding flats: he'd add a little bit of sugar to the seeds before he sowed them. The sugar shows up against the black dirt so you can see where the seeds land. Lobelia seeds, however, are tinier than even the sugar crystals. When I add a little sugar to the cup of seeds to "see where they hit," the sugar crystals stick out of the Lobelia seeds like boulders in sand.

But if you mix the tiny seeds with a half-teaspoon of sugar, that helps the seeds scatter out, it keeps them from clumping too much, and you can see where the seeds and sugar fall into the soil. You will have to keep mixing the sugar and seeds as you sow, because the two substances will not stay evenly mixed. Fine sand also works with tiny seeds. You can use this method whether you're sowing the seeds scattershot or in neat rows.

If you prefer to plant seeds in rows, take a #2 pencil, dab it lightly into water and wipe it off, then touch the pencil to the seed to pick it up. Then you can use the pencil as your dibble – that is, use the pencil to make the hole and plant the seed.

Or pour some of the seeds in one hand over your soil. With the other, dust them off your hand onto the soil. You should get good coverage this way, unless you have chapped skin or sweaty palms. Or just shake them carefully from the corner of the packet, or take a pinch and sprinkle them around.

I think that the scattershot method of sowing works best in a greenhouse that needs to produce lots of plants – more plants per tray, less room taken up in the greenhouse, fewer trays to have to wash later. Either method is fine; it's a free country.

Press the seeds into the soil (though watering them will do that for you), then cover the seeds with soil according to the package instructions. Nearly all the seeds will need soil on top. Seeds that need light to germinate will need just a sprinkling. Once they germinate, sprinkle enough soil over them to cover the roots.

In the greenhouse, I used a super-fine spray head on my garden hose to water the seedlings, taking care not to let the water puddle in the tray. If you can find a watering can with a super-fine spray head on it, you're set.

Damping-off disease

Damping-off can be a real problem in a seeding operation. Damping-off is a fungal disease that causes seedlings to keel over right at the soil line. Fungus loves damp soils and humidity.

To keep from contending with damping-off disease, keep the air circulating around the seedlings, especially when you get a lot of cloudy days in a row. One year I had a lot of cloudy days, and the greenhouse was cool, so when I watered the trays the moisture would not evaporate out of the soil. The humidity in the greenhouse was super-high. Conditions were perfect for damping-off disease, and that's what I got.

To protect against this, give the seedlings as much sun as possible, and keep a fan running over the trays to keep the air circulating. If you have damping-off, you may have to throw the infected tray out, because if you see the fungus in one part of the tray, the fungus has already infected the rest of the soil. If you want to try and salvage the remaining seedlings, sprinkle Actinovate, a fungicide, over the soil. Or use cinnamon, or finely ground sphagnum moss. (I just take dried sphagnum moss and rub it between my hands over the top of the seeding tray.) Also, spraying chamomile tea, using one tea bag in two cups water, works nicely (and it smells good). These methods change the pH of the soil's surface, making it just acidic enough to discourage the fungus. A small change in pH is often enough to discourage fungus – a good rule to keep in mind when dealing with other kinds of fungus in the garden (blackspot, for instance).

Once the seedlings come up, turn the tray occasionally so the seedlings don't all end up leaning hard to the right when you transplant them in the garden.

Don't be cowed by all these guidelines. Not everything is going to come out perfectly. Gardening is a field that requires the cooperation of living things, and living things can be bullheaded and contrary. If you're a gardener, you must be willing to wing it! You can always make adjustments next year.

Cold Frame Setup

I have the simplest cold frame setup in the world – several bales of hay against the side of the house with an old storm window on top. I suppose I could make a simple box with scrap lumber, but, to quote Pig from the Kipper cartoons that my daughter likes to watch, "Oh, I'm not at all clever like that."

Once the cold frame is set up and the ground inside is warm, mix some potting soil into the top layer of the ground, then sow lettuce, spinach, carrots, radishes, and salad greens. You might have mixed success with the root vegetables, but it never hurts to try. (I only say this because carrots hate me. I can't get those things to grow worth a damn. Beets, on the other hand, are my friends.)

Keep a thermometer inside the cold frame, and one outside. Open the cold frame a crack when daytime temperatures rise over 40 degrees, and wider when temps hit 60 degrees. Of course, be sure to close the cold frame at night.

Cats and other animals can sometimes be a problem in the cold frame. When I opened my cold frame to let it air out, I'd come back and find a cat lounging right on top of my seedlings in the nice warmth and going, "Ah, this is the life," until I chased him out of there. After that, I would open the cold frame just a crack – enough for hot air to escape, but not enough for an indolent cat to squeeze in, the jerk.

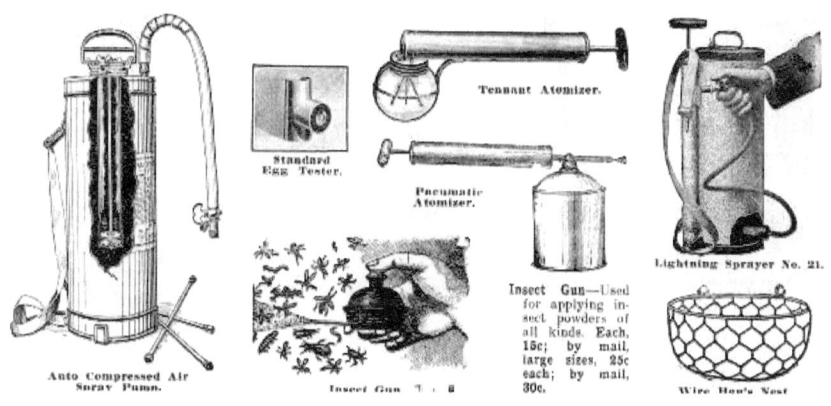

Spring Supply List

Winter is a good time to compile a supply list. Go through your fertilizers and tools and whatnot and see what needs fixed, what needs replaced, and what you need to get. It's better to do this now, than to get right up to a perfect day for sowing or seeding or spraying or fertilizing and have to waste that day running to the garden center. Well, a trip to the garden center isn't a complete waste of the day, but you know what I mean.

First on the list this year is a little gallon sprayer and a half-pint of insecticidal soap concentrate. These are for any bug infestations that might start eating my crops. Squash bugs are particularly nasty. You can spray insecticidal soap on your crops right up to the day before harvest with no adverse effects, though you'd still want to wash the soapy taste off the vegetables. Insecticidal soap is non-toxic, too. However, some plants don't take too well to insecticidal soap – it makes spots on the leaves. Check the label before you buy to be sure that all your plants are okay with it.

Neem oil is also a good, non-toxic insecticide, as is pyrethrum, a low-toxicity pesticide extracted from *Chrysanthemum cinerariifolium*, a type of daisy. Do not get pyrethrums mixed up with pyrethroids, the synthetic form, which is toxic. (Whatever you plan to spray, always read and follow label direction. More insecticide is not better, it is simply more – and the extra ends up in our groundwater supply.)

Floating row covers can be a valuable asset in the garden. They let sun and rain in, but can keep a light frost at bay. Best of all, they also keep invading insects out. When you put your vegetable plants in the ground, lay the floating row covers over them and peg them down to the ground. When squash bugs and flea beetles show up to lay their eggs on your plants, they are out of luck! And they can't eat your plants to shreds or spread their diseases from plant to plant. (You might still have to deal with cutworms, which come up through the soil.) When the flowers appear, uncover your plants so bees can get in to pollinate them – the squash bugs and flea beetles are mostly finished with laying eggs by then.

One drawback is that you can't get in to weed under a floating row cover – so mulch them!

If you need a compost bin but don't really want to spend a hundred bucks on some compost turner, a homemade compost bin will work just fine. Wire together three pallets, set them next to the garden, and pile all your compostables in there. Add a touch of country elegance by growing some clematis up the sides. Or use hay bales instead of pallets, because these will rot away inside and add to your compost, while keeping it toasty warm in winter.

To add nourishment to the ground, grab some bone meal, some blood meal, and some Bradfield fertilizer, which is all-natural and contains alfalfa. Alfalfa releases a chemical called triacontanol, which gives plants a growth boost. And of course keep that compost handy.

Don't leave the bone meal or blood meal unattended outside because dogs and cats will eat it. One time I was planting perennials and had a bag of bone meal nearby – I'd dig the hole, toss in a small handful of bone meal, dig it into the soil, and then plant my flowers. I went in to get a glass of water. When I came back out, Brownie, the neighborhood dog, was gnawing on the bag like it was a nice ribeye steak. A lot tastier than Kibbles n' Bits, apparently.

Invest in a soaker hose for this year's garden. These hoses, made of recycled tires, "sweat" water along their length. Lay the hose at the foot of your plants, cover it with mulch, attach the end to your spigot, turn it on, and you're in business. A soaker hose waters your plants slowly and deeply, doesn't waste water (it all goes directly to the roots), and, once you've laid it in your garden, doesn't require any effort besides attaching the hose and turning on the spigot. Just don't forget to turn it off at night!

Do be careful with the hoe around a soaker hose, because one careless swipe will puncture the hose. But if this does happen, you can easily splice it back together with a hose repair kit, available at any hardware store. When I used soaker hoses in the garden, I needed a way to get water from the spigot clear out to the soaker hoses. I had an old leaky garden hose lying around, so I cut several lengths of hose, then attached male or female parts, depending on what part of the soaker hose I needed to link to. These connected the length of soaker hose from the spigot. That kept the water off the sidewalks, where it didn't belong.

Soaker hoses work better than a sprinkler in the garden. (Do keep the sprinkler for your lawn and the kiddos, however.) Watering through the sprinkler is good for washing off dusty, dry plants, but a lot of water ends up evaporating instead of going to thirsty plant roots, so it's not a sustainable alternative.

For cats that leave presents in the mulch ... I was thinking about planting catnip at the far end of the yard as a trap crop, hoping that might divert their attention. Then again, I've known cats who will sit right next to a catnip plant and not care a bit. These guys are contrary as hell sometimes. What might work better would be to invest in three small Super-Soaker water guns, one for each window that looks over the garden. Anytime I glimpse a cat sniffing around the mulch, I'll stealthily raise the window and *fire!*

It might be as ineffective as the catnip patch, but it would be a lot more fun.

February List o' Things to Do!

* **Start seeds of slow-growing plants** this month – onions, leeks, celeriac, and celery – under lights to get a head start. Other seeds: broccoli, cauliflower, Brussels sprouts, and cabbage. All of these will be transplanted to the garden later.

* **Try raising a crop of leaf lettuce** under lights as well.

* **You can start Early Girl tomatoes** (or any early tomato variety) under lights, then, when the weather is mild enough, transplant to the garden with a Wall o' Waters to help protect the plants against all the frosts.

* **Take a chance and sow peas,** lettuce, spinach, and radishes under cover in the garden for early crops, if the ground isn't frozen. You might be taking a chance here with these crops – they might not grow – but faint heart never won fair garden. In other words, you don't win if you don't play!

* **Don't work garden soil** if it's wet or frozen. Squeeze some of it in your hand. If it is sticky, don't till or dig. If you can't even dig any of it up because chips of ice fly everywhere when you try, you had probably better wait a while.

* **If you know** you're going to have a lot more seeds than you need for your garden, team up with a neighbor or some friends to exchange garden seeds. This might even help you find some new varieties that will become a perennial favorite (pun totally intended).

MARCH

March has begun, finally. I love this time of year. It's like Schubert said in his song "Withered Flowers:" "Come on up, all you flowers, for it won't be winter forever and May will indeed be coming."

It's that time of year when people stand at potting tables in greenhouses, potting up bare-root plants as the big oil heater kicks on behind them. It's that time when garden centers make that final inventory of supplies and plants they need, and start hiring nursery help as the customers come surging in, all wanting tomato plants (which you shouldn't plant outside until May 1). It's the season of hope, when you dream of how great your garden will be and you can't wait to get your hands back in the dirt. The little scillas and crocuses pop up, the maples bloom (and the spring allergies start again), the grass starts greening up, and after a warm rain, and the world smells like rich leafmold and soil.

People start raking up their yards and burning yard waste (where ordinance allows). The lines at the car wash and Dairy Queen are long. Garage sale signs start popping out on every telephone pole.

Spring is such a good season.

Turn Your Weeds into Green Manure

If you have a lot of winter weeds growing all over your garden, and you don't want to pull them up, then till them under! Winter weeds can benefit your garden by being a valuable green manure that will break down and add nutrients to the soil. A green manure is basically a way to fertilize the soil by tilling green plants into the ground so they can break down and add nutrients to the soil. Tilling them directly in might be slow work, because henbit and chickweed tends to wind themselves around the tines, and you will have to stop the tiller now and then to unwind them. It might help to first mow the weeds down, and then till them in while they're still green. Just be sure to till the weeds under very thoroughly so they don't start growing back.

If you use your weeds as green manure, **till them under at least a month before you put out your seeds.** This will give your green manure time to break down completely. However, if you're just going to skip seed planting and put in the tomato and squash plants in May, then knock yourself out. There's enough time for it to work!

Green manures improve the soil structure, because the decomposing plants add ingredients to the soil that help to bind together soil particles. The result: a more crumbly soil that's easier to till.

Me Versus the Tiller

The area where I planned to put the flower/vegetable garden had been grass-free for about a month when I finally borrowed the tiller from our neighbor Rose. As I dragged the tiller out of her shed, my dog Marcus nosed the box of bone meal, and I remembered I needed to get soil amendments. So the dog and I went to the farm store in town. I got nine bags of manure and two of bone meal. When I got into the cab of the truck, Marcus glued his nose to the bone meal. I made a mental note not to leave it sitting around the yard where he might eat it.

Marcus and the cat, Boots, both awesome guys.

We poured the bagged manure on the ground and my daughter ran through the dirt piles as I raked them out. Then I poured out the bone meal, Marcus supervising intently.

I checked the soil to be sure it was dry enough for tilling. When you till or dig in wet soil, you're tearing up the structure of the soil. Soil that was loose and crumbly will turn into solid clods, or will form a water-repelling crust on the surface.

When in doubt, take a handful of soil and squeeze it. If water drips out, forget it! save your tilling for another day. If you open your hand and the clod is still stuck together due to the moisture in it, better wait on tilling. But, if you open your hand and the clod falls apart in little crumbles, then grab the tiller and go to work.

My husband started the tiller for me. Marcus challenged it briefly until we got him out of the way, then I started tilling. Or tried to. The tiller wanted to walk along the surface of the soil instead of digging in. I leaned back, throwing my weight into trying to get it to work. Unfortunately, the tiller has a hundred pounds on me.

My husband watched the tiller drag me around the yard for a while, then said, "How about I help you run that?"

He put on his headphones, turned down the throttle, and had the tiller chew a hole into the ground. Then he let it walk out until the wheels got stuck in the hole, and he let it chew another hole. After each hole was dug, he'd haul the tiller backward through the soil to chew up any ground it had missed. He sang Clint Black songs at the top of his lungs as he worked.

When he'd finished two rows, he turned the tiller over to me. Now the work was easier because I'd figured out (from watching The Man) that I needed to let the tiller work instead of fighting it. I kicked the throttle up to high. As the tiller dug, I'd ease it to the left, then to the right to get any compacted earth I'd missed along the sides, and also to get rid of that untilled strip right in the middle, underneath the machine. Then I'd lift up just a little on the handles to help the tiller dig into the ground.

It took several hours overall, but finally I walked the tiller back into its shed and I was done. Mr. and Mrs. Robin started cleaning worms out of the garden before I had it put away.

Marcus lost interest in the bone meal after it was tilled in. Whew.

Now all that was left was raking out all the grass clods, and smoothing out the garden, and setting up all the gardening plans, and planting all the seeds, and all the watering and weeding and picking and maintenance, and did I mention weeding? for the whole rest of the year. Piece of cake!

Other Tilling Hints and Helps

Tilling is one way to prepare the soil for spring – or even for fall. There are a few things you can do to make your tiller (and your soil) work more for you.

Before you till, be sure to renew the organic matter in the soil. This is an important part of your gardening to-do list every year. Soil microorganisms and worms are constantly breaking down the organic matter for your plants, and you've got to keep adding more to replace all that gets used up.

Great sources of organic material include leafmold, compost, or rotted manure – so spread these ingredients over the ground and till them in.

Tilling is a good way to prepare the soil for spring. But there are a couple of reasons why frequent tilling can be detrimental for the soil. One minor problem is that tilling can burn organic material quickly. The air that the tiller incorporates into the soil causes the microorganisms to work faster to break organic material down.

Tilling destroys earthworm burrows, putting a dent in the earthworm population. A healthy soil has a high population of worms, so when you kill some of your worms this way, you could be shooting yourself in the foot. Earthworms are a big help in many soils – night crawlers will bring nutrients from deep in the subsoil up to the surface. Little worms will create airways around plant roots, aerating the soil, and their castings are pockets of pure fertilizer.

To avoid this, you might till once a year. Also, leaving some areas of the garden untilled (for example, by including a strawberry or asparagus patch inside the garden) will give the worms a place to survive and multiply.

You might even leave small portions of the garden fallow instead of tilling. Plant some alfalfa, clover, or other legume to fix nitrogen for a year, especially if you know you are going to be busy this year. Then next year, till the area under and plant your corn there. This is a good option if you know that you're going to have extraordinary circumstances – such as a new baby, a major surgery, a prolonged trip – that will make it difficult for you to keep up a garden for part of the year.

If there's no area of your garden that you can leave fallow, then interplant clover among your vegetables. Use clover to cover your garden walkways, or plant white clover between your rows of corn. Legumes add nitrogen to the soil, bring in beneficial insects, and act as a living mulch through the heat of the summer. While other gardens are withering, your garden is nice and cool with all that green on the ground.

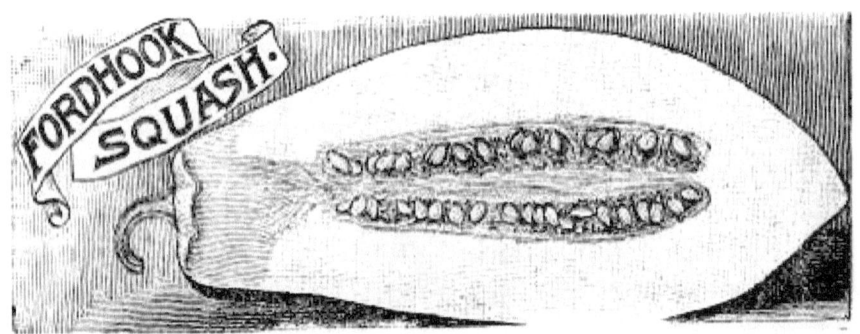

Seeding, Planting, Transplanting Outside

For seeds you want to plant directly in the garden this spring, you can get an early start by making a seed tape. Make a sticky paste of flour and water. Then tear strips of newspaper 1 inch wide. Put a thin layer of the flour paste on the newspaper, then (using a pencil) space the tiny seeds on the paper. Set the tapes aside and let them dry, then store. This spring, you can put the newspaper right into the garden and put a little soil on top. The newspaper will decay and the seeds will come right up.

For transplants, you'll need to take a little time to harden off the seedling. That is, they'll need to be acclimated to the temperature outside for a little while. One way is to take them outside every day for a little while. If you have a cold frame, you have a good place to keep the seedlings protected outside until they can get used to the cold days. Every day, put them outside for a little longer each time.

Hardening off seedlings usually takes a week or so. Then transplant them outside and keep an eye on them for the first week or so. Once they get watered into the ground, they should be all right, though it would be a good idea to cover them when the night temperature is forecast to drop close to freezing.

Ways to Discourage Rabbits and Squirrels

Rabbits and squirrels are cute, but these animals can be nuisances in the garden when they eat all your lettuce and repeatedly dig up your seedlings. (Is there any way we could train these animals to dig up weeds instead?) There are several ways to deal with them, and these tips can be used for any other critters that are wreaking havoc in your garden.

Repellents
If the rabbits and squirrels eat the plants in the garden, try repelling them with mothballs, blood meal, or ground limestone. Or try sprinkling cayenne pepper on the plants. Be careful sprinkling this so you don't start sneezing your lungs out. These will need to be replaced after a rain.

Gadgets with motion sensors are also available. When a critter ambles by, it will squirt a stream of water to chase it off. Kids love these.

It helps to have a cat or a dog that likes to chase things.

Electric fences
You might also set up a low-voltage electric fence low to the ground around your garden or yard. One strand of wire should work fine. An electric fence can be run off the AC current from

your house; others can be run on batteries, which would provide less of a zap.

Don't use red transformers on the fence, though, since hummingbirds are attracted to them. Also, be sure to keep all the grass trimmed around the fence. Grass or weeds touching the wires can short it out.

With an electric fence, it would probably be a good idea to purchase a voltage tester to help fix it when it shorts out, as well as a warning mechanism that will let you know if the fence is on or off so you don't have to test it yourself. "Is this thing working?" ZAP. "Yep, it is!"

If the rabbits squeeze under the fence because their fur insulates them from the shock, then wrap a little piece of aluminum foil on the fence and put peanut butter on it. When the rabbits touch it with their little wet noses, they'll learn to stay away!

Other fences
You can put a regular, uncharged chicken-wire fence around the garden, but you'll have to bury it a foot into the ground so the rabbits don't dig underneath. The squirrels will still be able to climb over the top.

If rabbits nibble on your young trees, wrap trunk guards around the trees. If you use wire, keep it far enough from the trunk so they can't press their noses through and reach the bark. Take the wire off when the tree gets too wide.

Live trapping is also an option. Rent traps from a rental store, bait them with carrots or apples, and then release the rabbits far away.

Shotgun

If you live in the country and the rabbits and squirrels are on your property, you can shoot them and serve them for supper, thus keeping down the population.

Shoot animals on your property only, though, and nobody else's. Otherwise you would be guilty of trespass and out-of-season hunting. This is also true of any trapping you may do, whether it's lethal or non-lethal.

Chickens will "help" you in the garden, or something.

WHEN DO I PLANT AND SEED IN MARCH?!

When I was over at Grandma Mary's, I saw a package of lettuce seeds (Black-Seeded Simpson) on top of the microwave. "Oh, it's that time of year again!" I said.

Time to clean up the garden and get out the tiller and put the tiller back because the soil's too wet and get the tiller out again and till it up and rake out the clods and then start agonizing about what vegetables to plant this year.

The planting dates I give below are for Zone 5, since that is where I live, and I'm a little Missouri-centric. If you are in Zone 5, these dates should work for you; in other places around the country, look to your local University Extension service for the dates that work for your area.

If you don't know where to find your local University Extension folks, just go to a computer with an internet connection, Google "University Extension" with the name of your state, and you will be in business. I can't praise these guys enough, because they're a good service and they're free and they have information about just about everything under the sun. Just give them a call, or stop by, or email them, and they'll fix you right up.

According to the Missouri University Extension Center, **beginning on March 25 you can plant these seeds directly in your garden:**

Beets, carrots, cabbage plants, plants, collards, garlic, kale, kohlrabi, head lettuces, mustard greens, onions (both seeds and sets), parsley, parsnips, peas (though cowpeas must wait until May), radishes, spinach, and turnips.

After March 25, these young plants can be hardened off and transplanted into the garden:
Broccoli, Brussels sprouts, Chinese cabbage and cauliflower transplants.

Though conventional Grandma wisdom states that potatoes can be planted on March 17, the University Extension Center recommends planting them on April 1. All I have to say about this raging controversy is that you have to find out what makes your potatoes in your garden happy. Perhaps you could plant part of your stock now and wait to plant the rest on April Fool's Day, and see what turns out best -- or even if you notice any difference between the two crops. Be sure to note results in your gardening journal for next year's planting.

March List o' Things to Do!

* **Local businesses will start displaying plants** for sale. Don't give in to temptation yet unless you have a cold frame or greenhouse.

* **If you have a cold frame,** start seeds for cole crops and other cool-weather crops. Plant seeds for carrots, radishes, lettuce, spinach, etc., directly in the soil in the cold frame for the earliest crops on the block. Don't start tomato plants unless you have a really big cold frame, or a supplemental season-extending device like water-walls or row covers.

* **Indoors, start seeds for tomatoes,** peppers, and eggplants under lights.

* **If the seedlings you started last month** are big enough, transplant them into pots, egg cartons, six-packs or four-packs, etc. How do you know they're big enough? The seedlings have

several sets of leaves, not just seed leaves (the two tiny leaves that are the first to open when the seed germinates). They look big enough to pick up and not break. Also, they look strong and healthy.

* **If the winter's been dry,** you might start watering your asparagus, raspberries, strawberries, and rhubarb, and other perennial plants. This will help them as they gear up for spring.

* **Wait to mulch your vegetable garden** until after the temperature has warmed up. Mulching cold ground will keep the ground cold, slowing the growth of your plants.

* **Plant asparagus and rhubarb roots (below)** as soon as the ground can be worked. Asparagus like a garden soil that's rich in rotted manure and compost. Next year, let the asparagus grow, and don't pick the new shoots (well, maybe you can pick a few) so the plant can get strong.

* **Before your till, put down fertilizer.** Use compost, rotted manure, or an organic fertilizer like Bradfield. One or two pounds of store-bought fertilizer per 100 square feet is plenty.

* **As the weather warms up,** start taking mulch off the strawberry patch so the plants can green up a little. Look out for weeds, though, because those are all ready to green up, too! Knock 'em down before they get too big.

* **Don't forget, a sudden frost** could turn all your seedlings into sad brown mushy things. Remember to bring tender plants in at night, or cover them with an old sheet or blanket.

* **Get the compost pile warmed up and going.** You've probably been putting scraps out here all winter – now is a good time to invigorate the pile if it's been slow. Stir in some manure to really get the pile warmed up. Throw in all those winter weeds for additional "green" materials, and throw in old, chopped-up leaves (a mulching mower is best at this operation) for "brown" materials.

APRIL

Planting Potatoes

Early April is time to be thinking about the almighty potato. Consider what varieties of potatoes you want, as well as where and how you want to plant them.

The old saw, "Plant your potatoes on St. Patrick's Day," won't work if the ground is wet or frozen. Better to wait for the soil to warm up.

There are a number of different potatoes available, such as Norland and Cobbler (with the earliest harvest dates); Viking, Pontiac, and Yukon Gold (with a mid-season harvest), as well as Kennebec and Mayfair (which round out the season).

Did you know that some heirloom varieties come in colors such as blue and violet? Blue potatoes include All Blue and Purple Peruvian. For ruby-red potatoes, get Red Thumb, Caribe, or Irish Treasure. Then you can make a patriotic potato salad. Most of these varieties can be ordered over the internet.

Purchase certified disease-free seed potatoes. Seed potatoes should be firm, not soft and wrinkly.

You can plant potatoes from the supermarket; however, these have been treated with a chemical to keep them from sprouting while they languish on the shelf. For best results, sprout before planting – that is, put them in a dark place for a little while until little buds or sprouts form on the eyes.

Cut the potatoes a day or two before planting, so the cuts will heal (below). This helps keep the potato from rotting in the cold, wet ground. Each potato chunk should have an eye or two. If the potato is egg-sized or smaller, plant it whole.

Choose a planting area with slightly acidic soil. Don't plant them next to a gravel road, or in a place that has recently been limed. Alkaline soils cause a disease called "scab," which creates rough spots on the potato. Keep the wood ashes away from the potatoes for the same reason – they are very alkaline.

Don't plant potatoes in a place where you have already planted eggplant, peppers, other potatoes or tomatoes, since all of these plants are in the same family and share diseases such as blight. Diseases can linger in the soil from one year to the next.

Dig a six-inch-deep trench, put a little compost at the bottom, space the seed potatoes a foot apart, then cover them. Don't worry about getting the exact spacing. In the good old days, Grandma would give us kids the sacks of cut-up potatoes. We dropped them into the trench as we walked, and someone else followed with a hoe and covered them. None of us carried a ruler. We harvested so many little red potatoes to mash up and eat with tons of butter and salt, and they all tasted great.

Keeping these rules of thumb in mind, 9 to 12 pounds of seed should plant about 100 feet of potatoes, if you drop them a foot apart.

Once the potato vines grow four to six inches tall, start mounding the soil around them. As they grow, keep mounding the soil higher until you have a mound about 5 inches tall. Congratulations, you have created a potato hill, allowing your tubers to grow in darkness, protected from the light. (Light creates green areas on the potato, which are mildly poisonous – and they taste bitter.) Once they're hilled, cover them with a nice mulch. This keeps the light off the potatoes, keeps in moisture, and keeps them cool.

If you have a lot of mulch at hand, you can plant potatoes above the ground in a 6- to 12-inch layer of hay, straw, or shredded leaves. An old bushel basket with the bottom busted out is great for keeping the mulch in one place. Add more mulch as the potatoes sprout. You may get up to 2 feet of mulch. Don't worry, the plants won't smother. When harvest time arrives, reach into the mulch and pull out the potatoes you need. Once the potatoes are harvested, spread out the mulch. Next year, you'll have a lot of good, rich soil in its place. This method makes for quick planting and clean potatoes at harvest.

Once the potato vines blossom, you can carefully dig into the side of the hill to start harvesting new potatoes. Take only what you need so the potato plant can go on producing. When harvesting new potatoes, remember that immature potatoes don't store as well as mature tubers.

When the potato vines die, it's time to harvest the mature spuds. Wait a week or two to harvest to allow the potato skins to toughen up, which makes them less vulnerable to bruising. Harvest on a dry day, wash the tubers off, and let them dry before you store them.

Then break out the sour cream and enjoy!

It's a Bug-Eat-Bug World

The beneficial insects page in a horticultural supply catalog is not for sissies. These pages, which list predatory insects that the organic gardener can use to control pests, can make the gardener feel like she's fallen into a Stephen King novel. These beneficial insects are vicious, devious ... and boy, are they hungry.

For instance, protozoans called *Nosema locustae* are used to control 90 percent of grasshopper and cricket pests. The protozoa are placed on bran flakes, which unsuspecting young grasshoppers eat. Once the protozoa infect these young grasshoppers, other grasshoppers will seek out these infected grasshoppers and cannibalize them. Then once these grasshoppers are infected, they get cannibalized, and so on. It's like a grasshopper *Walking Dead.*

Or perhaps you'd prefer some fly parasites, also known as trichogrammatid wasps. The female wasp lays her eggs inside the pest fly pupae. When the eggs hatch, the tiny wasp larvae consume the fly pupae from the inside.

I apologize to those of you who, upon reading that, spit out your coffee.

Cicada killer wasps, and other wasps, do the same thing. Cicada killers, which are huge wasps (but they aren't mean and are actually kind of fun to play with) will dig burrows going three to five feet into the ground. At the end she builds a little chamber. Then she goes out, catches some cicadas, stings them to immobilize them, and drags them, still alive, into the chamber. She'll lay an egg on them and seal up the chamber – and when the wasp egg hatches, the larvae has plenty of food to devour.

The insect world is cruel and vicious, which gardeners can use to their advantage. Organic means of pest control allows gardeners to control pests safely and effectively, without having to use harsh chemicals that kill or adversely affect the living things in the area. If you have an insect pest, it's best to seek an organic solution.

Assassin bugs will capture other bugs by jabbing them with their long proboscis – a strawlike eating tube. The assassin bug then travels around with their proboscis stuck in the other insect like they're carrying around a Slurpee from the 7-11.

Gardeners can buy "killer snails" (that would be a cool name for a rock group) that feed on pest snails and slugs. They can get ladybug larvae that eat live aphids, and praying mantises and pirate bugs that chow down on any insect pest that gets in their way.

Dragonflies and damselflies are good allies in the garden. Sometimes you might see a damselfly – a small dragonfly like a blue darning needle – flying around the garden and repeatedly bashing her head against the back of a leaf. Look closer. She's actually grabbing whiteflies off the back of the leaf. Dragonflies and damselflies also catch bugs while skimming through the air, and munch them while flying.

Ladybugs (above) are big sellers in organic gardening catalogs – kids love them, and so do grown-ups. When you release a bag of ladybugs in the garden, the adults will eat bug pests, but eventually they'll lay eggs and then leave. After a while you'll see tiny, black and orange larvae that look like miniature Komodo dragons on your garden plants. These are ladybug larvae, and they will eat aphids and other insect pests.

Praying mantises are like tiny lions, devouring anything that crosses their paths. They'll often be sitting in a garden with the remains of some insect clutched in its arms, chomping away. This ultra-predatory tendency leads to problems when a cluster of mantis eggs hatch – the tiny mantises will grab their brothers and sisters and eat them!

So the world of insects is filled with bug gore. E.B. White acknowledged this when he wrote *Charlotte's Web* (the only children's book to feature a pig, a rat and a spider as heroes). Wilbur, when he meets Charlotte the spider, is heartbroken to find that his new friend is so bloodthirsty.

"I have to get my own living," Charlotte tells him. "I have to be sharp and clever, lest I go hungry."

Thus do a spider's wits -- indeed, any beneficial insect's wits -- makes for our gain in the organic garden.

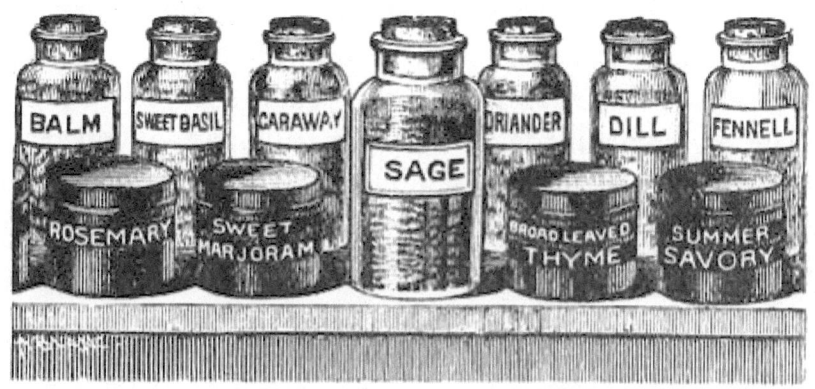

SEEDS OF POT AND SWEET HERBS.

Herbs in the Garden

There's always so much emphasis on color, form, and texture in the garden. When you think about it, however, a garden should have more to it than meets the eye. Fragrance gives special touch to the most utilitarian garden.

It's neat to combine scents the way you would combine colors – gathering a bouquet of scents through the plants one selects for such a garden.

That's why herbs are such an asset, whether in the vegetable garden, in a formal knot garden, or in the anything-goes-yet-it-looks-so-beautiful English garden. Herbs add color, scent, and functionality if you need fresh thyme and sage for the roast.

Many herbs are grown for their foliage. Some, such as tricolor sage, have variegated leaves that sport a cream edge. Lamb's ear, considered an herb, has soft, silvery leaves. In fact, some gardeners cut off the blossom stalks of lamb's ear, because after the plant blossoms, the leaves become sad and bedraggled. The blossoms aren't very showy, at any rate, but the leaves can really put on a show. It's a great nectar plant for honeybees and bumblebees. Try not to let it go to seed, as it gets very weedy.

Santolina, also called lavender cotton, is a silver-leaved herb that is used in knot gardens in the same way boxwood is – as a tidy shrub that can be kept trimmed to edge the garden.

If you like hummingbirds, grow pineapple sage. This plant has a sweet pineapple smell. I roam everywhere in order to find and purchase one of these, just so I can pluck the leaves in my garden at my leisure. Though it's an annual, this sweet-scented herb can grow into a small shrub, up to four feet in good soil, and bears brilliant red flowers that hummingbirds love.

Herbs come in so many fragrances. Thyme can smell like lemon, while mints and basil can smell like chocolate. That's another neat thing about herbs – their versatility.

Herbs can add color and scent to the vegetable garden as well. Here, you can harvest entire branches without messing up the overall appeal of your garden. I just go out and snip a little basil and the yellow squash I'm cooking tastes so good because of it.

My daughter loves to pick the leaves off the tricolor sage plant and put them in her pocket. "Leave some leaves on the plant," I tell her. "Okay," she says, then she picks off some more leaves. I've tried to interest her in the other herbs – I'm wild about basil at the moment – but she ignores them and makes a beeline for the tricolor sage.

Interplanting herbs with your vegetables also breaks up the monoculture of the garden. Nature hates a monoculture – where one species dominates – so she sends bugs, diseases, and weeds to break up the monopoly, allowing other species to compete. Interplanting herbs with large vegetable crops breaks up the monoculture and attracts beneficial insects.

When harvesting herbs, bear a few things in mind. Gather the leaves or stems before the herbs flower, and keep flowers trimmed off – clipping off the flower stalks keeps flavor in the herb leaves. Also, the best time to pick herbs is early in the morning.

Herbs can be kept for about a week in the refrigerator in a plastic bag with holes punched in it. For longer storage, dry the herbs in the microwave. Place four or five stems on a double thickness of paper towels; cover them with a single paper towel. Then microwave them on high for about two to three minutes, until the leaves are brittle.

Herbs can be frozen, too. Lay them out on a tray and put them in the freezer until they're solid. Then package them up, label them (ALWAYS label them!), and get them back in the freezer.

Early April Planting Guide

These dates are all for Zone 5. If you are in a different zone, you can find more information on these planting dates by visiting your local University Extension center. I'm using the Missouri University Extension guide for the dates below, and these dates should be good for Zone 5.

If you need your local planting dates because you do NOT live in Zone 5 (sorry), then call, email, or stop by your local Extension office, and they'll give you all the information you'll ever need right away.

In early April ....
Plant asparagus (be sure you have the bed prepared for this perennial vegetable), beets, cabbage, cauliflower, Chinese cabbage, endive, lettuce, mustard greens, New Zealand spinach, parsley, parsnips, peas, potatoes, radishes, spinach, and turnips.

The window is closing for planting broccoli, Brussels sprouts, carrots, collards, kale, kohlrabi, onions, rhubarb crowns, and head lettuce, so get them in the ground as soon as you can.

Plant these after April 10:
Parsley, parsnips, rhubarb crowns, and Swiss chard (on the 15th).

It's time to plant spinach, turnip greens, carrots, cabbage, and asparagus. Hurry to finish planting the carrots, peas, kale, and broccoli, since the windows for planting these are nearly closed. Hold off on sweet potatoes, winter squash, tomatoes, peppers, green beans, and watermelon until May.

After April 25:
You can plant bush beans (but not bush limas), your second crop of lettuce, mustard, radishes, rhubarb, Swiss chard, and turnips.

By now you should have planted beets, broccoli, Brussels sprouts, cabbages, carrots, cauliflower, cabbage, collards, endive, kale, kohlrabi, head lettuces, New Zealand spinach, onion bulbs, parsley, parsnips, peas (but not cowpeas yet), potatoes, and spinach. Whew!

Get the last of your broccoli, Brussels sprouts, cabbage, and cauliflower plants transplanted into the garden before the weather heats up. Use a fertilizer high in phosphorus (or, ahem, compost) to get the little plants off to a good start.

The next big planting wave starts on May 1. Are you ready?

April Showers Means May Flowers ... and Weeds!

After an early April rainstorm, everything greens up beautifully – including the weeds. Pounce on them while they're still small. Put the pre-emergent down on the lawn, pull up the weeds in the flowerbeds, and hoe out the vegetable garden.

By catching the weeds while they're still young and tender, you save yourself a lot of work (which you can expend elsewhere in the garden – April and May are busy, busy months)! Also, a clean garden is a great morale-booster when it's time to plant.

April Planting Tips

It's time to plant corn, beans, and cowpeas. However, cucumbers, eggplants, muskmelons, okra, peppers, pumpkins, squash, sweet potatoes, watermelon, and the all-wonderful tomatoes should wait until May 10-15 for planting. None of this applies to gardeners with an elaborate covering system prepared against surprise frosts.

When buying tomatoes, look for the letters A, F, V, T, or N on the tag. These tell you what diseases this particular tomato variety is resistant to.

An "A" means it's resistant to Alternaria stem canker. An "F" or "V" shows resistance to Fusarium or Verticillium wilt. "T" means the tomato is resistant to tobacco mosaic virus. (Incidentally, cigarettes can bring this disease into the garden soil, through the tobacco in the cigarette. Even the cigarette ash can spread the virus, which is not destroyed by that little fire from your match.)

Finally, an "N" means it's resistant to nematodes, tiny white roundworms that burrow into the roots.

Plant tomatoes as deeply as you can. Roots will form along the stem that's underground. That will give you a deep-rooted plant that will withstand drought.

April List o' Things to Do!

* **You vegetable producers** probably have your cole crops in the ground already, as well as your onion sets, and can survey with pride your rows of seed all planted. I envy your efficiency. You've been planting your raspberry and blackberry bushes and the grapevines, too. But hold off on your tomatoes! We could still have a cold snap!

* **When you plan** where you want to put tomatoes, peppers, and eggplant, don't plant them in the same places as last year. All three of these plants are in the same family – the Nightshade family – and share the same diseases in common. By rotating planting areas, you avoid a buildup of pests and diseases in the same spot.

That's something to keep in mind when planting other plants, too. For example, when marigolds are planted in the same bed again and again, you start seeing nodes on their roots when you pull them up in the fall. The nodes, which look like large bumps, indicate that nematodes have been building up in the soil. These tiny white worms dig into the roots (which cause nodes) and sap the life of the plants. Swap out the marigolds for a different plant – one that is not susceptible to nematodes – and the nematodes will die off.

Variety is a good thing.

* **Some folks use marigolds** as trap plants for nematodes. They plant them in the garden, give the nematodes time to dig into the roots, then yank them (and the nematodes) out of the soil. Bye-bye buggers!

* **Seed a second crop of lettuce** and peas outside (provided, of course, that you've seeded a first crop of lettuce and peas earlier).

* **Start cucumber, summer squash,** watermelon, and cantaloupe seeds inside. Also plant luffa and hard-shell gourd seeds inside, but be sure to soak these seeds overnight before planting. Hurry up, summer!

* **Toads eat slugs,** potato beetles and cutworms, among other things, so they can be an asset to your garden. Encourage toads to come to your garden by building a cool, moist spot for them to live during the day. A board or a flat rock with a space underneath does well, and a partially submerged clay pot would

be a suitable house. Once a toad finds a home, it may become a permanent fixture in your garden – that's especially good if you have slugs hiding out in your vegetables.

* **Use plastic film to warm up the soil** where you want to plant warm-season vegetables next month.

* **Need some quick plant markers in** the garden? Cut up white plastic milk jugs and make your own. Write on them with a permanent pen or marker.

* **Grow some sweet potato slips for planting in May.** Start slips just like you did in grade school when you were learning about how to grow plants – cut a sweet potato in half, stick toothpicks in the sides, then suspend it in a glass of water that keeps the bottom end submerged. Shoots will sprout out of every eye. When they reach 6 to 9 inches long, slip the sprouts off the potato (that's why they call them slips) and stick them in a different glass of water to root them, or plant them in a little soil. Here in about six weeks, after all danger of frost is past, you'll be planting these puppies outside.

* **Attract ladybugs and lacewings (below),** both of which attack garden pests, by spraying a sugar-water solution around your garden. The sugar solution resembles the honeydew secreted by pests such as aphids, which attracts ladybugs. Mix five ounces of sugar into a quart of water, then spray the water over the places where you want the ladybugs to go.

* **Plant some herbs next to your kitchen door** for fragrance and for cooking. Give your sage and basil plenty of room to grow. Keep the low-growing thyme and chives at the front of the border, and put rosemary in a pot to bring inside over winter. Protip: If you grow mint or oregano, keep them in containers. These plants spread like crazy. When you plant them directly in the ground, these mints will take over your yard and garden, then come into the house and eat stuff out of your fridge just to be rude.

* **Start harvesting from asparagus** patches that are three years old or older. Pick sparingly from two-year-old plants, and nothing at all from asparagus planted last year.

* **Grass clippings are a boon in the vegetable garden** – a quick, easy mulch that is good for your vegetables. Of course, make sure that the clippings do not contain herbicidal residues, which can kill the crops. Even clippings that were gathered three mowings after the herbicide was applied can be toxic.

* **Planting melons this year?** Choose varieties that are resistant to diseases. Then, build hills several feet apart in a place where they receive full sun. Mix an inch of compost into the hills, then put two or three plants in each hill. Feed the seedlings two weeks after you set them out, then twice more during the season, and make sure they receive an inch of water every week. After the fruits form, pay extra attention to watering, but stop fertilizing at this time. About three weeks before the harvest, cut back on watering the melons – too much water may cause the fruit to crack.

* **Start harvesting asparagus and rhubarb.** If your rhubarb plant tries to bloom, cut off the flowering stalks. Keep up with the weeds around the plants. Pick and crush asparagus beetles crawling around on your fronds (like this guy below).

* **Get cages not just for tomatoes,** but also for peppers and eggplants. Use the 3-foot tomato cages for the peppers and eggplants. The tomatoes may prefer the 6-foot tall rings, because a lot of them just grow through the tops of the 3-foot rings and then flop onto the ground. Put stakes around the cages in preparation for stormy winds or overloaded plants.

* **Plant some early tomatoes** if you're brave! Have your Wall o' Waters handy, or some blankets, for cold snaps.

* **Go ahead and plant an early sowing** of warm-season crops such as beans, squash, sweet corn, cucumbers, and New Zealand spinach. If they freeze, oh well, just replant 'em. If they survive and thrive, well now you have an extra-early crop! Gardening is like gambling, only cheaper, AND you get vegetables.

MAY

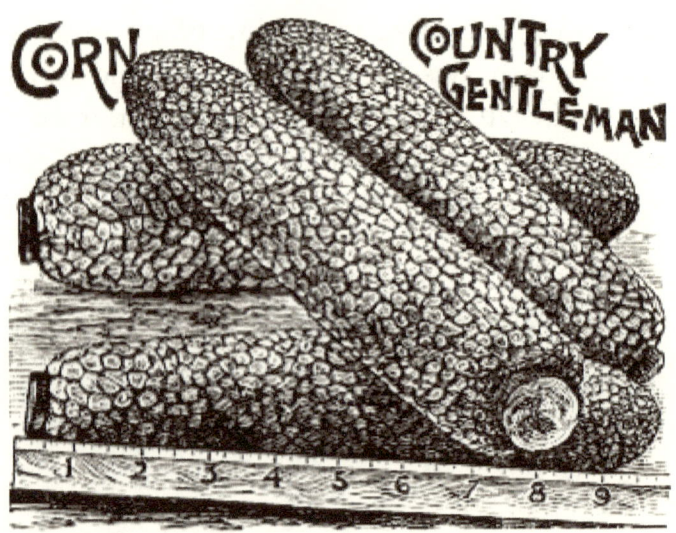

A Quick Planting Guide to Corn

Oh corn! So nice to have in the garden! But so demanding! "Give me good soil! GIVE ME ROOM," it keeps saying. Then again, the end product is pretty worthwhile, so what the hey.

First, the space. For a great corn crop you'll need deep, well-drained, fertile soil. The fall before you plant, add compost or well-rotted manure and till it in. You might even consider growing a cover crop of legumes, like red or white clover, in the place where you plan to plant your corn in fall. When winter comes, cut down the legumes right at soil level, let them wilt for a few days where they lay, and then till them in. That's called a **green manure**. The green stuff will be incorporated into the soil over the winter, and in the spring the soil will be fertile and ready to go.

In northwest Missouri (Zone 5), corn can be planted on May 15, though these days I think we can safely move that date back to May 1 due to global warming. At any rate, for the quickest germination, the soil should be at 65 degrees. Corn in this warm soil should germinate between 4 to 7 days.

Plant corn seed one inch deep, four to six inches apart. Rows should be about a yard apart. If you have hybrid corn, plant it in blocks 4 rows wide so you get great pollination and full ears of corn. When grown in a block, corn silk will more readily catch the pollen that drifts down from its tassels. Without good pollination, you get tiny ears with no kernels.

Plant the seed more thickly if the seeds are not treated with fungicide (treated seeds are a pink color). The fungicide helps keep the kernels from rotting in the soil.

When choosing corn varieties to plant, take care not to mix hybrid corn varieties, because they will cross-pollinate, which will do weird things to the flavor of your kernels. To avoid cross-pollination, keep your plots of corn 25 feet apart – OR plant the different varieties two weeks apart to keep pollination dates from overlapping. So for example, you might put the sweet corn in one patch, the super-sweet corn in a different patch, and the popcorn elsewhere, all far apart, to prevent cross-pollination.

Open-pollinated varieties, however, are safe from all this rigmarole.

Corn needs a good soil with lots of nutrients, especially nitrogen. Till compost in deeply when preparing the bed so the feeder roots will be encouraged to reach deep.

To extend your corn harvest, plant some early, middle, and late-season corn. Or have several plantings several weeks apart. So, make a planting with an early corn variety. Then, two weeks later, plant another block of early corn, and a block of mid-season corn. Then, later, another block of mid-season corn. Then a block of late-season corn. However you prefer to do this, keep planting and staggering the plantings until June, depending on how long your planting season runs.

Here's how to figure out your last planting date for your corn crop. Look on the back of your corn packets to see how long the corn takes to grow from planting to harvest – then get a calendar and count the days backward from your frost date. The date where you end up will be your last planting date.

The corn will need to be fertilized as it grows. Apply a high-nitrogen fertilizer twice during the season: first, when corn is 8 inches tall, and again when the stalks grow their tassels.

When cultivating around your corn, be careful of the feeder roots near the soil's surface. It might help to hill up soil around the base of the corn plant to keep weeds down and to also give the corn a better hold in the soil. Also, once the corn has grown a little and the soil has warmed up, throw down a layer of mulch to keep weeds down, to cool the soil in the hot summer sun, and to keep the soil moist. Straw is great for this purpose.

Leave suckers on the plant. Removing them actually seems to reduce yields.

Look out for corn earworms (below). If you see some damage on your ears – your CORN ears, not your own ears – like a little bug has been boring into them, carefully open the husk and dig out the little caterpillar and squash it. Don't let it fall to the ground or it will pupate and pop out of the ground next spring as a moth to lay eggs in next year's crop.

Corn earworm!!

Remove and burn old stalks in the fall to kill off any insect hosts.

If your corn gets smut, a fungal disease, you'll see kernels and sometimes tassels start growing monstrous gray boils. Remove and destroy affected plants before the boils burst and spread the fungus spores to the rest of your crop.

Smut!!

Harvest the corn as soon as the silk turns brown and dry, when the tips of the ears are round but not hard. Check by puncturing a kernel with your fingernail – if the corn is ripe, you'll see a little milk come out. (This is how my grandmas check corn at the store to be sure it's ripe.)

After you have harvested your corn, shred the stalks and compost them, or burn them if your compost pile is not hot enough to kill off any pests or diseases that the stalk might have collected.

Tips on Growing Melons

Melons should go into the ground about the same time you plant your tomatoes, beans, and cucumbers. If you're sowing melon seeds, make a little hill of rich soil into which you've dug some compost. At the top, plant three to five seeds about two inches apart and about an inch deep. Once the melon plants have developed two sets of leaves, pull up the smallest and weakest vines, leaving the two strongest plants.

For an earlier start, you could lay a sheet of black plastic over the soil and the melon hill, make a little hole in the middle, plant the seeds through the hole, and let the melons grow. The black plastic will warm up the soil for the plants, serves as a mulch for them (certainly it will help keep weeds down), and makes cleaning up the melon mess at the end of the season much, much easier! Of course, anchor the plastic so the melon plant won't sail away over your house during the next thunderstorm.

Melons like about an inch of water every week (actually this rule applies to most every garden plant). If you use drip irrigation or a soaker hose in your garden, you can bring water directly to the ground next to the roots, keeping it under your mulch or black plastic. It also prevents fungal diseases in the leaves, because direct irrigation keeps the leaves dry.

Fertilize with a well-balanced fertilizer, such as 5-5-5, and add several inches of compost, or a straw mulch, or a mulch of grass clippings or shredded leaves. (If you are using the black plastic, mulch won't be necessary.)

If you see an early burst of flowers but you don't see any tiny fruits following the flowers, don't panic. The earliest flowers that show up on the plant are male flowers – pollen-bearers – and won't set fruit. Male flowers are small, appear in groups, and have no fruit at the base. Female flowers, when they develop, are larger, and they will have a tiny vegetable-shaped bulb at the base of the flower.

If for some reason the bees are scarce, or if you want to be absolutely sure that every flower gets pollinated, you can pollinate the flowers yourself. (P.S. This goes for any of the cucurbits – pumpkins, squash, cucumbers, cantaloupe, and watermelons.) Dip a paintbrush or a cotton swab into the anthers of the male flower – the parts that are covered with golden pollen – and make sure you see the little yellow specks of pollen on it. Then dust the pollen onto the pistil in the middle of the female flower and make sure it sticks there. Once the female flower is pollinated, the flower will close and fall off, and the fruit will develop.

PEAR-SHAPED GOURD.

Vertical Gardening: Put Your Crops in the Air

I've set up three sturdy poles about three feet apart and tied them together at the top to grow my pole beans. Melons and peas turn out great when grown on trellises. The "Sakata Sweet," a baseball-sized yellow melon should take to a trellis with no problem. Cantaloupes, honeydew, and cucumbers can be grown this way, too. This gets the fruit off the ground and up in the air where you can harvest it more easily.

If you have visions of your trellis-grown cantaloupe breaking prematurely off the vine and smashing on the ground, you might start collecting old pantyhose or onion sack webbing to make little hammocks for the melons. If you go that route, you might also amuse yourself by giving your cantaloupe a canopy to keep off the sun and a tiny tropical drink with a paper umbrella in it. Then, this summer while you're weeding in the sweltering sun, you can rest assured that at least *somebody* is having fun in the garden.

Mexican Bean Beetles

In May, the Mexican bean beetles are a nuisance in the bean crop. The adult beetles are large, yellow-orange ladybugs with small spots. Their larvae look like spiny orange Dr. Seuss characters. Beetle and larva both cause some real damage, seeking out your snap and lima beans and defoliating them, and turning the leaves into yellow-green lace. The adult beetles overwinter in leaf litter, then emerge in spring to wreak their havoc and also to lay rafts of bright yellow eggs on the undersides of the bean leaves.

Spined soldier bug getting ready to snack on Mexican bean beetle larvae

Handpick the beetles as you pick the beans, dropping the beetles into a jar of soapy water. (See if the kids will help you with this.) If the infestation is really bad, spraying the plants with soapy water will help kill them off. Another thing that might help is to pull up and destroy an old bean row as soon as a new one starts producing. Then beetles, larvae, and eggs are destroyed.

Or, plant early-maturing bean varieties, and bring the harvest in before the bean beetles do too much damage.

May's Favorite Pickling

Mulch Down The Weeds

If you have weeds in the garden, then get them managed quickly. Hot weather plus lots of spring rain equals weed heaven. For you it means a lousy time. Who wants to work in 90 degree heat? Not me!

Get out there and get the weeds cleaned up, as quickly as you can. Once you get an area cleaned up, throw the mulch right down on it to keep the weeds down.
Mulch is a lifesaver. It keeps the rain from splashing dirt all over your plants. It keeps the soil moist so you don't have to water as often. It keeps weeds down, keeps the soil cool, and makes everything look so nice. Mulch is the gift that keeps on giving.

You can even mulch right over the weeds and smother them. I became fed up with hoeing my garden, especially with the heat and humidity coming on. So, I got a big pile of newspapers and laid entire sections around the plants, overlapping them right over the weeds. (You CAN use glossy pages from the newspaper – pretty much everything is printed with soy ink these days, which is just fine for the soil.) There should be at least five pages in each section. If you lay down the sections while they're still folded, you get ten pages. Then I piled hay on top and watered the whole thing down. It really works!

With small plants, you can tear newspapers into strips so they'll fit between your plants, then cover the newspapers with grass clippings or wood chips. (Add extra nitrogen to the soil if you use wood chips – they will pull nitrogen out of the soil as they decay.) This will keep the weeds down until the seedlings get big enough to fend for themselves.

Just be sure to cover the newspapers completely with a thick layer of mulch so the next thunderstorm won't blow the pages around. Also, I've discovered that when some of the newspapers showed from under the mulch, my cat would shred the exposed pages. But she's weird.

If you have chickens and let them roam in your garden to scratch around the vegetables, then newspaper mulch might be out of the question. In some cases, you might be able to hold the newspapers in place with rocks placed through the mulch. Or maybe not. Chickens are notoriously single-minded about scratching up the very places you don't want them scratching up. Kind of like cats.

Stop the Grasshopper Invasion

When the plagues of grasshoppers start, it seems that nothing will stop them outside of dynamite. However, when you start working on killing off the grasshoppers in May, when they're small, you have a greater chance of slowing down the onslaught and protecting your vegetables and plants.

Get 'em when they're small
Grasshoppers are easiest to kill when they're less than ½ inch long. Start early and start looking for grasshopper hatching grounds. Good places to look for hatching grounds are ditches, fence rows, pastures, or roadsides. If there is a large number of small grasshoppers in one area, you're in the right place. These young grasshoppers are called nymphs, and look just like full-

sized grasshoppers except they are ½ to ¾ inch long. When in the nymph stage, grasshoppers don't move very far from where they've hatched. If you find hatching grounds in your area, flag them, spray them, and check back on them later.

You can also use *Nosema locustae* bait at the hatching grounds. *Nosema locustae* is a protozoan microbe that causes diseases in grasshoppers. The protozoans are mixed in a bait that the grasshoppers eat – it's available as Nolo Bait and as Semaspore. Infected grasshoppers are cannibalized by non-infected grasshoppers, which in turn get cannibalized in one of those delightful "Nature Gets Gross" turnabouts.

If you use the bait correctly, while the hoppers are still small, you can get a 30 to 40 percent kill rate. However, Nolo Bait isn't as effective on mature grasshoppers. Also, spread the bait when the weather is mild for best results. But when you set out the bait in the hatching grounds, it does work. Also, broadcast some bran bait between the hatching grounds and your garden to catch the young hoppers moving in. Nolo Bait works best if you use it correctly for several years in a row. The disease will carry over into the next year.

Protip: Every insect pest, weed, or disease is most effectively dealt with when they're small, or just getting established.

How to control grasshoppers when they're big

Once the grasshoppers get big enough to start flying, usually about late June, then you're in more trouble, because now they get hard to eradicate. Insecticides and sprays still work; however, when you kill the grasshoppers in the garden, more hoppers will migrate in to take their place. It seems you could spray all summer and not get anywhere.

To keep the grasshoppers from entering your yard, use a trap crop to catch the grasshoppers as they travel in from surrounding pastureland. Grow a tall, lush stand of grass around your yard. Fertilize it with nitrogen to make the grass really succulent and tasty, and water it. The grasshoppers will latch on to the grass and start eating that. Then you can spray the grasshoppers inside the trap crop as they eat.

One note: don't let the trap crop of grass dry out, and don't mow it, because then the grasshoppers will head to the next oasis of green: your garden.

Other means of grasshopper destruction
Organic insecticides such as insecticidal soap or neem oil will work against the grasshoppers, but only if you hit the actual insect with the spray. Floating row covers also protect your garden against grasshopper onslaughts. However, if you have crops such as cucumbers or squash under them, the plants will have to be hand-pollinated in order to bear fruit, since the cloth also keeps bees out. Also, if the grasshopper infestation is really bad, they may eat their way through the cloth! In that case, you may have to use metal window screening to keep them out. Or maybe dynamite.

Desperate times take desperate measures
If you live in the country, get some poultry. Guinea hens are best, but if you like peace and quiet, you'd better settle for some chickens or ducks.

Some birds also feed on grasshoppers, including bluebirds, mockingbirds, brown thrashers, crows, and sparrows. Attract birds to the yard with a clean birdbath, good habitat, and some hiding places.

Often, hand-picking the hoppers can help, especially if you need fish bait.

The situation gets more serious if the grasshoppers are moving in from surrounding pastures that have had no treatment, especially once the pastures start browning in the summer heat. And most treatment beyond hand-picking is ineffective against flying adults. Sometimes I cut them in half with my garden shears. Once, the back half of a hopper jumped about five feet into the air after I cut it in two. It would have been really interesting if I hadn't been so busy gagging at the sight.

Spotted Cucumber Beetles are Evil

Okay, maybe they're not evil, but they sure aren't doing you any favors.

I usually call them green ladybugs (mainly because people instantly know what I'm talking about when I call them that), but these are actually spotted cucumber beetles. These eat the leaves of cucumber plants and other cucurbits (like muskmelon and pumpkins). They can carry bacterial wilt from plant to plant while feeding, and this disease causes your plant to wilt, first in individual leaves, then nearby leaves, then through the whole plant. Sometimes the plant can succumb to the disease in only seven days.

If your plant is wilting and you think it might be bacterial wilt, try a simple test to be sure. Cut the stem and squeeze both cut parts. A sticky sap will ooze from it. If you stick the cut parts back together, then slowly draw them apart, you'll see ropes of sap stretch between them if the plant has bacterial wilt. In those ropes of sap are millions of bacteria.

Spotted cucumber beetles affect other plants. I usually find these beetles eating the stamens and petals of white or yellow roses. Then all the petals fall off prematurely. They'll also get all over your pole and bush beans, and destroy your peas.

It's possible that these beetles also spread squash mosaic virus and fusarium wilt. I wouldn't put it past them, to be honest.

Whenever I see one of these beetles, I just grab it in my fingers and squish it. If you don't want bug guts on your fingers, carry a spray bottle of insecticidal soap and spray the beetles when you see them. You might put floating row covers over the cucurbits (though watermelon is generally not affected) to keep the beetles off.

To keep numbers down next spring, till your garden in the fall. This may help expose beetle larvae and kill them. Clean up the garden and burn all cucurbit debris. Mulch your plants to deter the insects from laying eggs around your plants and slow larval migration through the soil. Another way to catch them is to put yellow plastic bowls filled with water and a little bit of dish soap near your plants. Cucumber beetles for some reason will crawl into the yellow bowls and drown in the dish soap water.

Tomatillos seem to be an effective trap plant for spotted cucumber beetle larvae. Let them grow in your garden until they start showing signs of larval damage – then pull them up and destroy them. The beauty part is that you can harvest fruit off the trap crops until you pull them up. They reseed so vigorously that you'll always have new plants popping up around the garden. If you are a tomatillo fan, then this solution may be just the thing for you!

May List o' Things to Do!

May is always a hectic month in the garden, so I'll jump right in.

* **Plant all seeds and plants** in your vegetable garden now: It's time!

* **If the young plants in your garden are six inches tall** or taller, start piling mulch around them to suppress the weeds that are already competing against them. Use grass clippings, chopped-up leaves, and compost – anything that will break down into the soil and add organic matter to keep the worms, and all the tiny critters in the soil, alive and thriving. (A healthy soil biomass breaks down organic matter into humus, making the nutrients in it available to plants, and also improves soil health in a thousand different ways.) Mulch also keeps the ground cool, and keep that crust from forming on top of the soil that sheds rain like a duck's back.

* **If you're tight on space** in the vegetable garden, try vertical gardening. Set up poles and twine to grow peas, pole beans, melons, cantaloupes, and tomatoes. (The tomatoes will need to be pruned to get them to grow this way, but it's worth the effort.) The vines will toughen up enough to hold cantaloupes aloft – though watermelons might be a different matter.

* **Thin your carrots and beets** (baby vegetables, yum!) to prevent overcrowding.

* **If you're growing cucumbers** to make pickles, don't forget to plant some dill and garlic.

* **If you've had trouble with cutworms** in the past, make collars for your young transplants. Take pieces of thin cardboard, wrap them around the stems of your plants at ground level, tape them, and push them a little into the ground.

* **Keep harvesting asparagus** and keep killing those asparagus beetles.

* **Start planting sweet corn** as soon as oak leaves are as big as a squirrel's ear.

* **Every time you get a space cleared** in the garden from harvesting early crops, throw down some seeds or plant some warm-season crops.

* **Put some netting over your newly-developing strawberries** before the birds figure out what you have there. You might have to suspend the netting over the plants, using small sticks or poles, to keep the birds at a distance. If the netting's directly on top of the plants, a robin might sidle up to the edge of the strawberry bed and start helping herself, though she wouldn't be able to get to the berries in the middle. Or maybe you'd be fine with sharing a few berries. You decide!

* **When you plant your squash and cucumber hills**, put a stake next to the seeds. Then, after the vines have eaten up half your garden, you will still be able to locate where the roots are. This makes it much easier to water them.

* **Pick caterpillars from your broccoli** and cabbage plants, spray them with insecticidal soap, or pour soapy water over them.

* **Put a screen or a floating row cover** over your lettuce to keep it from bolting (i.e. going to flower, which makes the lettuce bitter) and keep your harvest going for a little while longer. This also keeps butterflies from laying eggs on them to unleash a thousand caterpillars on them. They flutter around outside the mesh, and if you get close enough, you can hear them cussing.

* **If you have trouble with slugs,** lay some boards on the ground. Every morning, look under the board and destroy any slugs you see there. (I generally throw the big ones into some remote section of the yard – show them what it's like to fly).

* **Plant sweet potatoes, too.** These are very sensitive to frost, so plant them late. Make a raised berm of soil, tuck the slips into the soil about a foot apart, and firm them in. Some people cut the roots off the slips (don't worry, the slips will root as soon as they hit the soil) while others leave them on.

Protip: Lay black plastic on the ground around the sweet potatoes. This will keep the ground nice and hot for these guys, and will also prevent the vines from rooting. When vines root, they direct their energy toward making potatoes in that new spot. As a result, at the end of the season you get a bunch of tiny potatoes instead of a bunch of big potatoes. If you don't have black plastic, keep lifting the vines up so they can't root.

* **No need to add extra fertilizers** when planting herbs – too much fertilizer could affect the pungency of the herbs. Compost is fine, though.

* **When transplanting plants** in peat pots, tear off the top inch of the pot before sticking it in the ground to keep water from wicking out of the soil.

* **The best time to transplant** outside is on a cloudy or overcast day, or late in the afternoon as the sun's on its way down. Or, use a floating row cover to cover the new plants and protect them from the sun's heat.

* **Tomatoes should be planted** with about 2/3 of their stem underground. This allows a deep root system, and roots will grow out of the stem to add to this system. If you aren't able to

dig this deep, then lay the plant on its side and plant it sideways. Just be sure it is at least five or six inches underground.

*** Protip: To keep squash borers (see picture)** from laying eggs in your squash plants, wrap the base of your plant in panty hose. This usually foils them.

Squash borers doing the work of the devil.

JUNE

June is that month when everything looks great. The plants are coming up nicely, the rows are still tidy and green, and the mulch is staying in place, mostly. You've already had some good harvests. The plants look just as happy as they could be and are busy blossoming or bearing tiny vegetables. You can sit outside in the evening and enjoy how the garden looks as the first fireflies start blinking over the yard, and you entertain yourself with visions of the garden looking much like this in August, even though, by August, the whole place will actually look blighted if it were not for all the hulking weeds.

June is that great in-between month until the heat cranks up and the weeds start catching up to you. Try to work in the cool of the day – in the morning and late afternoon, if possible – to keep out of the sun and make your work easier.

Don't Become a Statistic: Avoid Heat Exhaustion

To work in the heat is no fun. I've stood in the garden with the sun burning my back and hair in 95 degree heat and the relative humidity at 95 percent – which means that when I drank my hot water from my water bottle (the ice had melted long, long ago), I would immediately sweat that water back out again – not that it did any good, since sweat ceases to be a cooling mechanism when it can't evaporate.

If you work out in the heat often, then there are a few things you simply have got to do to take care of yourself. Please, don't make yourself a statistic to heat exhaustion or heat stroke.

About heat-related illness
Heat-related illness is caused by three things: dehydration, losing electrolytes, and the body not being able to cool itself. When temps climb above 95 degrees, the body can't lose heat through radiation (that is, your body can't radiate out its heat) so losing heat through evaporation is the only option. But when the humidity is close to 100%, then evaporation doesn't work, either!

According to the National Oceanic and Atmospheric Administration (NOAA), about 175 to 200 people every year die from heat. During severe heat waves, the number could rise as high as 1,500. Fast treatment of heat-related illness is necessary; one is more likely to die if the heat illness has been untreated for two hours. The elderly are the most at risk from heat, as well as people who have heart disease, obesity, or skin problems that make it difficult for them to sweat. People taking medications such as beta-blockers, antihistamines, diuretics, and certain kinds of anti-depressant medicines should also take it easy in the heat.

How to beat the heat

First, wear loose, light-colored clothing. Cotton is good because it wicks out the sweat. A wide-brimmed straw hat is wonderful; find one wide enough to shade your neck. Be sure to drink lots of fluids, even if you don't think you need it. Sports drinks are best, but water works just fine, too.

Be sure to take breaks, too. If the temperature is 95 degrees and the humidity is over 90 percent, you're just not going to get as much done, so you have got to allow for more time. It's better to work before 10 a.m. and after 3 p.m., but even during safer times, it can be a real bear to work outside.

Heat exhaustion, which is non-lethal, can sneak up on you, so watch out for symptoms such as severe thirst, irritability, fast and shallow breathing, weakness, headaches, muscle cramps, nausea, and vomiting. Sometimes the heart may seem to skip a beat, or the skin may feel cool and clammy.

Heatstroke, which can be lethal, has most of the above symptoms, except the skin is hot and flushed. Also, heatstroke includes odd behavior that stems from the central nervous system shorting out – things like confusion, disorientation, unresponsiveness, seizures, or hallucinations. At its worst, one falls into a coma.

Note that, contrary to popular opinion, patients with heatstroke may still be sweating, though more often the skin is dry. Their temperature is over 104 degrees. They can't keep fluids down, they have severe abdominal pain. If you or somebody you're working with has these symptoms, *call 911 immediately!* This is a life-threatening condition.

If you or somebody you're working with is affected by a heat-related illness, do the following:

Get them cool. Put them in the shade, get them into the air-conditioning. If the person is alert, get them into some cool bath water; if not, sponge them or douse them with water so it doesn't get into their face. Get a fan to blow cool air over them.

Have them lie down. Elevate their feet slightly. Loosen or remove clothing.
Give them drinks. Sports drinks that replace lost electrolytes are best. Have them take small sips at first. Don't give them anything with alcohol or caffeine.
If the patient vomits, have him lie on his side.

IF THIS IS HEATSTROKE, CALL 911. The fastest treatment has the best chance of recovery. While waiting for the ambulance, do all of the above. Use ice packs, if possible, or bags of frozen vegetables, anything you can to bring the body's core temperature down.

Be safe out there.

Potato Blight Also Affects Tomatoes

Last year my Celebrity tomatoes developed dark blotches when the tomatoes were still very small and green. Each time the blotches quickly spread over the whole fruit and turned it into brown mush. I was aggravated. I found one tomato that I thought had made it to maturity, but when I picked it up, it turned out that the side it had been sitting on had been completely eaten away by rot.

This wasn't limited to the tomatoes touching the ground, either; tomatoes up in the air also had these blotches. And, even stranger, the tomatoes on the other side of the garden were doing just fine – no weird blotches, no rot. What was going on?!

My tomato had late blight, the same disease that caused the great potato famine in Ireland. (Potatoes and tomatoes are in the same plant family, the Solanaceae, so they share many diseases.) It develops in wet, humid, cool conditions – the kind of weather that my tomato plants had been experiencing. When you have a blighted tomato, one side of the fruit may look lovely, but the other side is slimy, rotted away, and brownish-black.

Late blight (also known as potato blight) is caused by *Phytophthora infestans*, an oomycete (a fungus-like microorganism). There are many different varieties and forms of blight, including anthracnose fruit rot (which makes bulls-eye shaped circles on tomatoes), early blight (where the petal ends on the tomatoes turn white with fungus), and Septoria leaf spot (which affects only leaves and stems, leaving them peppered with little black spots).

The leaves on plants affected by blight have little brown spots on them, as if a few drops of acid fell on them. On the back of the leaves, or on their tops, is a lot of white, powdery mold.

Blighted tomatoes.

It should be noted that in some places, the fungus has developed resistance to systemic fungicides. However, external fungicides – those that repel fungus by changing the pH of the outside of the plant – still do the trick if you start spraying them immediately after you plant your tomatoes and potatoes. Keep spraying the fungicide weekly (or as often as the label specifies).

The drawback is that the fungicide will have to be reapplied every time it rains. Also, by the time blight has appeared on your tomatoes, it will be too late for a fungicide to do any good. Fungicide can be used only as a preventative measure, not a cure.

However, some experts are not sure how much good spraying actually does, so you might take notes about your spraying program. See what works and what doesn't.

One way of outsmarting the disease may be to plant a wide variety of disease-resistant tomato plants, not just one variety. Even if you lose one or two tomato plant to the disease, the other varieties could resist it. Also, space the tomato plants well apart from each other. My tomato plants were scattered around the garden, well apart from each other, and this probably saved some of my plants.

Mulch the potatoes and tomatoes to avoid water splash-up from the soil.

Sometimes blight spores are splashed up onto the plant from the bare soil. Also, water early in the day so the plant has time to dry off before nightfall. Don't plant tomatoes where you had any potatoes, tomatoes, eggplants, or peppers the previous year. This can be tough if you have a small garden and not much room to move things around. Do your best.

Stem collapse from late blight.

Give potato and tomato plants plenty of space in the garden for good air circulation. You might trim off parts of your tomato plant, especially if it's a monster plant, to let the air travel through.

As soon as you see infected fruits or foliage, remove them and burn them. (Don't compost them unless your pile gets really, really hot.) Bag them up and get them out of the garden so no more fungus spores can spread. (Spores come out of the fruiting bodies in the brown spots.)

Some tomato plants, such as Mountain Fresh, Mountain Supreme, and Plum Dandy, show resistance to the disease. Look specifically for these kinds of tomatoes if you've had blight in the past.

When you fertilize, keep the nitrogen levels low. Nitrogen makes leaves juicier and more succulent for diseases. Don't save seeds from infected crops, since the disease can be spread through the seeds.

If blight gets out of control, you'll have to pull up and destroy any infected plants as soon as you see them. Keep spraying fungicide on the survivors.

If you had trouble with late blight in your garden last year, take preventative measures by spraying those vegetables with Bordeaux mixture every two weeks during hot and humid weather. It helps if you pick off the affected leaves, too, and dispose of them away from the garden (not in the compost heap!). Don't allow volunteer potatoes or tomatoes to grow in your garden.

Kill the Right Bug

One day at the greenhouse, we were scratching our heads over what caused the holes in the leaves in several flats of plants. "Shoot, it looks like slugs to me," I said. I made that diagnosis mainly because the plants were in a cool, moist place, partially shaded, and I figured there were plenty of places for slugs to hide.

So Rhonda got the slug bait while I began to clean out the flat, looking for the slugs on the undersides of the leaves and the pots. Nothing. I was getting puzzled until, as I groomed the plants, a soft green caterpillar dropped into my hand. A few plants later, I found another caterpillar clinging to a stem. Hey, the slugs were framed; here's the real culprit!

The moral: it can be difficult to know which insect is making those holes in your leaves or toppling your plant, since pest insects stay out of sight, hiding under leaves or feeding after dark. There are plenty of other insects in your garden who take no interest in your leaves but simply happen to be hanging around the injured plant when you come outside. These poor innocent bugs end up being the target of your wrath, decimated in a cloud of poison. Then the pests come out of hiding, unscathed, and continue to use your plants as a salad bar.

Here, as always, observation is the key. Take care to investigate a plant for pests by cleaning up any debris around the plant, sifting through the mulch, and looking under leaves. When you know for certain what pest to target, you buy the right materials to take care of it, instead of wasting money buying, say, slug bait for a caterpillar problem.

Often, the problem is solved by handpicking the offending insects and squishing them. If the thought of handpicking makes you squeamish, I have heard of gardeners using a handheld vacuum cleaner to snap up bugs. No need to spend $11 on a quart of chemical you might spray only once or twice and then have to dispose of.

Slugs and snails are pretty easy to catch. You look for moist conditions and rainbow slime trails. Diatomaceous earth dusted on the ground around plants will kill slugs by razoring open their protective slime coating when they slide over it. Diatomaceous earth must be replaced after a rain.

You can also try setting out shallow containers of beer for slugs, though this may not be a good idea if you already have a raccoon problem. For a non-alcoholic slug bait, and to keep drunken raccoons from trashing your yard, use three teaspoons of yeast dissolved in a cup of warm water.

Vine borers can be a problem on squash vines. These are caterpillars that chew their way into the vine several inches above the ground, and kill off the squash when it blossoms. If you notice a sawdust-like substance on the stem, cut into the injured place and dig out the borer. Then pile up soil above the injured place. If you've caught the borer in time, the squash will send out new roots and survive.

Another insect to look out for is the large, dark-brown squash bug. Try laying out scrap boards next to your plants. Every morning, turn the boards over and smash the squash bugs you find there. You can also find and smash the rafts of copper-colored egg clusters on leaves.

Squash Bugs

Squash bugs look like alien invaders from some Hollywood B-film. The adult bugs are grey or brown, and their backs look like a shield. They are often misidentified as stink bugs, and they are out to get your squash, watermelon, and cucumber vines. A large number of squash bugs feeding on the plants can cause the vines to wilt badly, but the wilt ends when you get rid of the bugs.

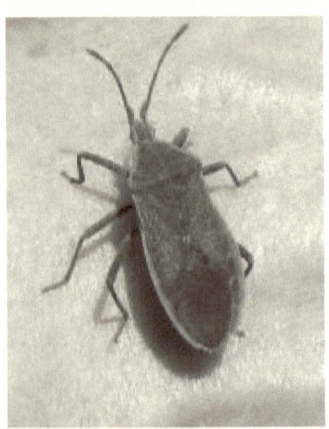

To catch squash bugs, lift leaves and crush the copper-colored egg clusters underneath. Clean up plant debris so the bugs have no hiding places. Lay several boards around the plants. Every morning, lift the boards and kill any bugs hiding underneath.

If the bugs are getting out of control, spray insecticidal soap or pyrethrum (an organic insecticide made from chrysanthemums), but this works best only when the bugs are still small. You have to spray pyrethrum directly on the pests, but they fall off the plant and die as soon as it hits them. Sometimes I carry a bottle with me so I can zap those green ladybugs in the rose blossoms, or the tobacco hornworms on the Nicotine flowers.

At the end of the season, destroy all vines and clean up the garden so adults have no place to spend the winter.

Next year, cover vines with a fine-mesh cloth to keep bugs out. Once flowering begins, remove the cloth so bees can pollinate the flowers. By then it will be too late for bugs to do their damage.

Fusarium, Bacterial, and Other Vegetable Wilts

Olin Cooper of Pickering, Mo., (who has since passed on, sadly) wrote, "I wonder what invaded my watermelons and muskmelons in the garden last year. Just as the melons started to ripen, the plants died and I did not get a melon from the entire patch. I realized that the same thing may happen this year, since the patch is very close to the same location as last year."

When possible, melons – cantaloupe, watermelons, honeydews, etc. – should be planted in a new place every year, because soil-borne diseases can survive in the same place. Also, some pests, such as root-knot nematodes, can infest the roots. Nematodes, or eelworms, are tiny white wormlike creatures that feed on roots.

I've noticed that they'll also attack marigolds that are planted in the same spot year after year. In fall, when you pull the marigolds up, the roots are covered with small, knotty galls. The roots of the melons will look the same way.

One bad disease is bacterial wilt. The vines wilt as if they are severely dry, and then they simply croak. Once the vines wilt, there's no chemical that'll bring them back.

When you put out young melon plants, keep an eye out for cucumber beetles. There are two types. Some of them look like green ladybugs – an article about them appears earlier in this book. Other cucumber beetles (in the picture below) are green with black stripes.

Striped cucumber beetles on a melon blossom.

Spotted or striped, these beetles spread the bacterial wilt disease from plant to plant. (The bacterium that causes wilt resides in the guts of some of these beetles. When they feed on your plants, they spread the bacterium into these open wounds. The bacterium then multiplies inside the plant, blocking the vascular system and causing the plants to wilt. Hence the name, bacterial wilt.) Zap those spotted cucumber beetles with insecticidal soap when you see them. You really have to move fast, because when you move to kill them, the beetles will immediately fly away or fall off the plant.

Another disease is Fusarium wilt. This is a form of the fungus that attacks tomato plants. Fusarium starts in the roots and move from there into the stems. If your plant is hardly growing and is all stunted, that's probably what the problem is. Then they wilt and die. Sometimes a white fungus starts growing on the dead vines.

If this happens, get the dead vines out of the garden ASAP so the spores don't spread back into the garden. Don't plant any melons or tomatoes there next year, or the next, because Fusarium will linger in the soil. Your best bet is to plant resistant varieties. Next year when you buy tomatoes or melons at the nursery, look on the tag under the plant's name for these letters – V, F, L, N – there. These denote what diseases the plants are resistant to.

For reference, here is a listing of what those letters mean:

A - Alternaria leaf spot
F - Fusarium wilt
FF - Race 1 & Race 2 Fusarium (lots of strains out there)
FFF - Fusarium, Races 1, 2, and 3
L - Septoria leaf spot
N - Nematodes
T - Tobacco mosaic virus
V - Verticillium wilt
St - Stemphylium (Gray Leaf Spot)
TSWV - Tomato Spotted Wilt Virus

Mainly it's the preventative measures that will save your crops: crop rotation from year to year, quick cleanup of diseased plants, and the merciless slaughter of those blasted cucumber beetles.

All my squash plants have succumbed to Fusarium wilt. So I have these big, tidy, mulched areas with nothing in them, though the Mr. Stripey tomatoes are already moving into the territory. We'll let them.

June List o' Things to Do!

* **Good eggplant production** means you need heat. Eggplants like temperatures between 80 and 90 degrees and plenty of water. Water well when the plants are young, then at least twice a week when it's hot and dry outside. Mulch always helps.

* **Flea beetles will make your plant leaves** look like they'd been shot full of holes (especially on the aforementioned eggplants). Spray the plants with insecticidal soap, or even get a bucket of mild soapy water and clean up the plants, getting the undersides of the leaves as much as possible. Then rinse the plants off with a hose. You might have to repeat this several times until all the flea beetles are hatched out (they'll keep laying eggs until you get rid of them, darn 'em).

* **Look out for crazy weather** and thunderstorms. If you have some new plants or transplants that you don't want to see destroyed, have a supply of plastic milk cartons with the bottoms cut out. Shove the milk cartons into the soil right over the transplants so the wind doesn't blow them away. Also, floating row covers are helpful if you can keep them well-anchored through the storm.

* **Watering will be crucial for the seedlings** and young plants. Soaker hoses, laid along the rows, work well. They simply drip into the soil, and they won't splash tiny seedlings out of the ground the way a hose will. Also, put down a nice mulch of

shredded leaves or grass clippings (only use herbicide-free clippings in the garden so you don't accidentally kill off your plants) in the garden. Mulch keeps the soil cool, keeps the soil from splashing up on the seedlings during a rain, holds moisture in, gives the worms and soil organisms something to eat, and is generally good for plants overall.

* **Try and keep up with succession planting.** If it's been a week or two since you planted your beans, plant another row. Do the same for corn. This will extend your harvest for a couple of weeks.

* **Put away your unused seeds** for next year. Seeds will survive for several seasons if you keep them in a tightly closed glass jar. Some people add a little packet of milk powder to keep the seeds dry. I don't know where you'd get that. Silica gel would also work.

* **Plant your pumpkins now** for Halloween harvest! Plant them as you would other melons, in hills, with plenty of space around them. Put a few handfuls of compost into the hill when you put the seeds in so you get healthier vines.

* **Stop harvesting asparagus** when the spears become thin, and give it a nice "meal" of compost or Bradfield fertilizer. Mulch them well so you're not fighting weeds there all summer.

* **When cucumber and squash vines** start to "run" and take over the garden, that's when you really need to step up your game against spotted and striped cucumber beetles (green ladybugs)

and vine borers. Have a bottle of insecticidal soap or pyrethrin and spray 'em when you see 'em.

* **Keep mounding up soil** around your potatoes to keep them from being exposed to the sun so they don't turn green and bitter.

* **To keep earworms out** of your roasting ears and your corn, apply several drops of mineral oil to the corn silk every three to seven days.

* **After the strawberry patch is all harvested,** mow the patch so the old plants are taken out and the young plants can grow up and get ready for next year. Thin out extra plants, and throw down some mulch to keep down the weeds.

* **If you've started Brussels sprouts seedlings** for a fall harvest, go head and transplant the baby plants outside when they're big enough. Also, seed broccoli, cabbage, and cauliflower inside. These can be transplanted outside later for a fall harvest as well.

* **Put a mailbox, mounted on a post,** in your garden for a weatherproof place to store small tools, labels, seeds, gloves, or other useful things you need as you do your gardening work.

JULY

Water your garden while wearing THE PUFFIEST SLEEVES IN THE UNIVERSE

Sometimes Plants Need a Break from the Heat

When the heat is particularly intense, plants will stop producing. Tomatoes, for instance, stop producing fruit when temperatures soar above 90 degrees. Cucurbits (zucchini, cucumbers, and muskmelon) might produce a bunch of male flowers but no female flowers, much to the gardener's frustration. (Unless she has more than enough zucchini.)

A number of vegetables are good with heat, including tomatoes, eggplant, cowpeas, lima beans, melons, and peppers. Sweet potatoes, okra, and southern peas (cowpeas) are the best at dealing with heat.

Even so, a number of these plants will stop producing fruit and might even drop their blossoms when temperatures stay above 90 degrees.

Watering the vegetables daily helps, and this is where soaker hoses are especially helpful. They keep the water down at the roots and minimize evaporation.

Mulch is also a great help in heat. Two to four inches of mulch will keep the sun off the soil. Once, when I was a city horticulturist, I was watering the dry, bare ground next to a plant and saw what I thought was dust coming off the ground. Then I realized that it was actually steam!

If the heat is especially brutal, or if the plants are wilting despite all your best efforts, lay some shade-producing materials over them – a shade cloth, old sheets, sheer curtains, or even a snow fence. A little shade can do the trick and pull the plants out of their doldrums.

When Weeds Get Out of Control

Sometimes you can't help it. You have to go on vacation, or a family emergency happens, or you have to attend a series of graduations – and you simply can't get out to the garden for several weeks. When you come back, it's weed city! And if you're in a place where you can't spend a whole lot of time weeding – or if the heat is too much – it helps to take emergency steps to get the situation under control.

If you have a bunch of weeds between the rows in your garden, use a string trimmer to cut them down. Don't beat yourself up if you clip off a couple of tomato or cucumber plants here or there. Just go after those weeds with a will. Scalp the ground if you need to, until you get the row cleared.

Or, if the row is wide enough, run the lawn mower straight up the middle of it! Then spray the weed stubble with horticultural vinegar, an organic herbicide, to kill it off. Again, be careful of your vegetable plants as you spray. Keep the applicator wand close to the plants you are trying to kill. This helps you control what you are killing. Don't spray on days with heavy winds, obviously.

Your goal here is to knock down a bunch of weeds before they get big or, worse, before they go to seed. Sometimes getting a big patch cleaned up can help you feel like you can tackle the rest of the mess – it's a great morale booster. Believe me, when the temperature is 95 degrees, you're going to need all the morale boosting you can get.

If the weeds are extremely bad, and if you know, for whatever reason, you're not able to handle the weeds, call a friend or a family member to give you a hand. Or, pay a lawn service (if they know what they're doing).

Another possibility is to simply haul out the tiller and just till it all under, weeds and plants and all, and start a fall garden! If it's too crazy, there's nothing like starting fresh.

Practical Harvest Tips

Now the vegetables are growing like gangbusters (the zucchini maybe a little too much so – some of the bigger ones have already learned to walk and talk and are now plotting the zucchinization of Planet Earth) – so now it's time to harvest!

Don't pick too far ahead of harvest. Sometimes the vegetables will ripen up (like tomatoes) but some will simply lose their fresh taste. Even the hybrid sweet corn will lose some of their sweetness if left out and unrefrigerated for 12 hours.

Cabbage can be picked at any stage – there's not much difference between the taste of cabbage in an old head or in a young one. Don't leave them on the plant too long, though, or their heads may split.

Snap beans should be harvested before the pods lose their tenderness. If they make a fine "snap!" when you break them in half, that's just right for eating. The bigger the pods get, the less tasty they can be.

If you harvest your broccoli when the heads are nice and compact (catch them just before they flower), you might get a second harvest of small heads right afterward. Cauliflower should be harvested when the heads are compact and white.

Try not to harvest while the plants are wet, because this allows diseases to pass easily from plant to plant.

Harvest the radishes when they're young and small. The taste turns sharp and bitter if you let them go too long.

Pick lettuce, peas, and corn early in the morning for the best taste and to keep it from wilting.

Wait for the ground to dry to harvest root crops such as potatoes and carrots. Mud could cause the roots to spoil.

To keep peas, beans, and cucumbers producing, harvest ALL the crops at once. Leaving over-mature veggies on the vine keeps new flowers from forming – all the energy goes toward seed production instead.

Tomatoes and cucumbers will keep longer if you leave a short stem on the fruit. Use pruning shears or scissors for best results.

Leave a half-inch of stem on top of your beets so they won't "bleed."

Be careful not to bruise your vegetables when harvesting them – this also causes spoilage down the road. Tie bunches of vegetables together with rubber bands. Washing your produce before storing it can shorten shelf life. Leave it dry.

Cure your onions, squash, sweet potatoes, pumpkins, and gourds in a warm, dry place for several weeks before you store them. If rain is in the forecast, cover them up.

Be sure to pick zucchini before it gets big enough to take over Planet Earth, thank you.

Bitter Melons and Other Asian Squashes

Asian squashes and cucurbits are excellent vegetables that have many culinary uses (and some non-culinary uses!) and some of them have excellent health benefits. Not many of these are seen in your run-of-the-mill seed catalogs, but if you go to a large farmer's market in summer, you will see a number of these squashes and some of their greens and blossoms for sale. The greens can be steamed, while the blossoms can be stuffed, steamed, and eaten (be sure to check the blossoms for bees before preparing them). Asian squashes can be cooked in so many ways, are easy to grow, and some of them have outstanding health benefits.

Bitter melon

Bitter melon is true to its name! A lady at the City Market was eating a bitter melon raw and snapped off a piece for me to try. I found it was indeed very bitter, but the flavor reminded me of raw string beans. Somebody at the City Market recommended that I slice the melon and soak it in salty water for about five minutes to remove the bitterness. I tried that before I cooked some bitter melon in scrambled eggs. Each bite of the melon released little juicy bursts of bitterness. I've heard that the bitter melons with smoother skin, the Chinese kind, are less bitter, while the wrinkly Indian kind are more bitter.

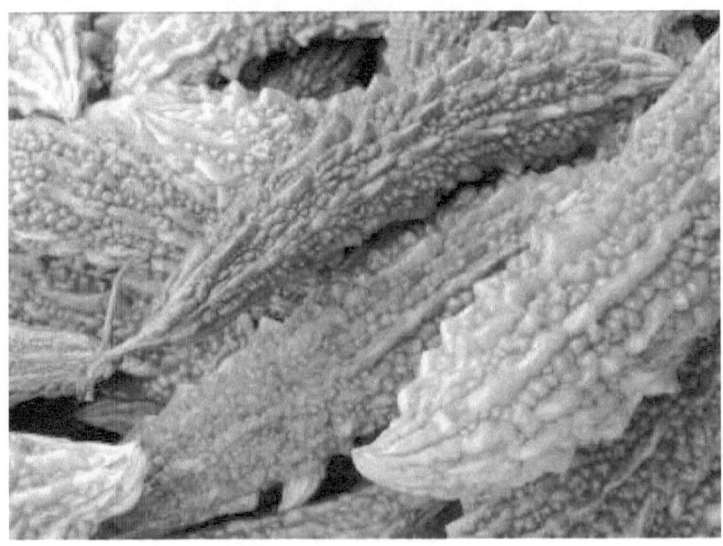

Bitter melon (*Momordica charantia*) can be stuffed with a mixture of ground pork then cooked in soup, or sauté the bitter melon with ground pork or lamb over rice, or in a stir fry with beef and black beans. Some cook it with sweet onions and spices in a dried curry. Some also slice bitter melon into thin chips, then fry them until crispy and sprinkle salt, turmeric, and chili powder on them. It is also made into a dish called pinakbet.

Recent research confirms what Chinese herbalists have known for centuries: that bitter melon can be used to treat type 2 diabetes. The melon contains four bioactive compounds that, when consumed, activate an enzyme that regulates metabolism and affects glucose uptake. One of the four compounds in the melon increased the oxidation of fatty acids and increased glucose disposal. Other trials demonstrated that bitter melon also helped insulin production in the pancreas.

Apparently a quarter of a fresh bitter melon is enough to help lower glucose in the body. Make it into a tea by slicing the quarter melon into about 1 cup of water. Heat the water to boiling, then simmer it for about 5 minutes. You can smash the fruit to get more of the juice out. Then take the saucepan off the heat, cover it, and let it sit for 10-15 minutes. Strain the broth into a cup, let it cool, and drink, and eat the soft melon.

I've found that, despite my wimpiness with bitter melon, I can drink the broth when I water it down and add a little instant chicken bouillon to it. I was too much of a wimp for even the boiled melon, but I can deal with the broth. I guess there is hope for me yet.

About ½ ounce (18 grams) of the dried fruit taken daily as a tea or in capsules also helps against diabetes. (If you dry the fruits, store them in halves or quarters to preserve their medicinal strength.)

Snake gourd
A lady at the City Market recommended that I drink the broth of the small snake gourds (*Trichosanthes anguina*) after boiling them until soft. The snake gourds I bought from her were small and cute, a dark green with white stripes the length of the vegetable, and no more than five or six inches long. When left on the trellis to grow, the snake gourd turns long, white, and curly and really does look startlingly like a snake, hence the name. The gourd becomes woodier as it's left to grow; I've seen only the young gourds for sale at the market. The gourds can grow up to six feet long.

Snake gourd at a market in India. To the right is some opo squash.

Snake gourd is a favorite in Indian cuisine, where it's also called padwal. In other places it's called tindora. The snake gourd has long grown in India, even being mentioned in some early Sanskrit writings.

Sauteed in butter, the vegetable tastes somewhat like zucchini. Snake gourd can be cooked in a vegetable soup, or battered and deep fried. A number of Indian recipes use snake gourd combined with beans or vegetables or in curries. This is a mild vegetable, and is peeled, split, the seeds removed, and cut up before cooking.

The snake melon needs a long growing season with long days, because the flowers open late in the afternoon. The flowers of the snake gourds are incredible. These white blossoms bloom at night, and delicate white threads curl from the petals, giving the flower an appearance of ghostly lace. Ghost flowers and snakes on a vine – what's not to like?

Opo squash
Lagenaria siceraria, which also goes by the names of bottle gourd, long melon, and Nam Tao, has several distinct varieties. One variety that grows in Africa has yellow flowers that resemble zucchini flowers and fruits that are also known as calabashes. The fruits are not eaten when mature but are allowed to dry on the vine and are used to make water containers, boxes, musical instruments, masks and other useful items. Many people in America make birdhouses from bottle gourds. However, an Asian variety has dainty white flowers and the long, narrow fruits that are a pale green; this variety, sometimes called the opo squash, is good for eating.

These long, light-green squashes have firm, white flesh, and can be sautéed like zucchini, stir-fried, or used in soups and stews. The opo squash are tastiest when picked young, because the fruit grows more bitter as it matures and starts to harden. The young shoots and leaves, which are soft and fuzzy compared to most squashes, can also be steamed and eaten.

This particular plant is, truly, older than Moses. According to archeologists Esquinas-Alcazar and Gulick, the bottle gourd was used in Peru around 12000 BC, in Thailand around 8000 BC, and in Zambia in 2000 BC. Such an incredibly ancient plant has definitely earned the right to be called an heirloom!

Silk squash, or luffa
The silk squash can be cooked and used in many different ways, which makes it popular in parts of Asia as well as in the Middle East and the eastern Mediterranean. Silk squash is also known as Chinese okra, sin qua, buab, and nethi beerakaya (*nethi* means ghee, which is a kind of clarified butter – the squash has a smooth, buttery texture when it's cooked.)

In America, the silk squash is grown for the luffa inside. The squash is left on the vine until it dries out. Then it's broken open, soaked in water for several days, and the inside of the squash is then taken out, dried, and used as a cosmetic bath scrubber. However, the young silk squash, if harvested when the fruit is still solid, makes a tasty vegetable. Very young squashes can be sliced and added to salads for their cucumber-like taste; they can be battered and deep-fried like okra, or sautéed like zucchini. The luffa can also be stir-fried with shrimp, or made into a chutney.

One variety of silk squash, *Luffa aegyptiaca*, is smooth-skinned, while another variety, called ridge gourd or angled luffa (*Luffa actuangula*) has longitudinal ridges; these ridges need to be peeled off before cooking because they make for tough eating!

This luffa has many of the same culinary uses as its cousin, but some people say that the angled luffa tastes the best.

Winter melon
This large cucurbit, whose scientific name is *Benincasa hispida*, has been grown in China since about 500 AD. This is a large, green melon with a thin coating of a white wax that keeps the melon protected against microorganisms, so it can be stored for a long time like a pumpkin or a winter squash. When properly stored, the winter melon, or wax melon, can stay fresh for up to a year without refrigeration.

Winter melon soup in summer is a good choice to help to cool down and detoxify the body. Winter melon soup is also good for a weight loss diet plan. The white flesh is somewhat bland when eaten raw, but is excellent in curries, stir-fry, and soup; winter melon soup is often served in the scooped-out melon, thus saving you from having to do dishes! The shoots, tendrils, and leaves may also be steamed for greens. Winter melons are also used in a candy called petha or in sweet pastries. The seeds can be fried and eaten like pumpkin seeds. The young fruits are fuzzy and of course are called fuzzy melon.

When making winter melon soup, scoop out the seeds and cut the melon flesh into cubes. These can be cooked with shrimp, crab meat, or pork.

So the next time you're at a farmer's market, or perusing an heirloom seed catalog, try out some of these Asian squashes, and experiment until you find a way of cooking them that you like. A lot of good surprises are in store.

THE BOSS. A distinct variety introduced a few years since. Shape as shown in accompanying illustration. Skin is black green in color. Flesh deeply scarlet, unusually sugary, crystalline and melting. The rind is very thin and tough. It ripens early, and is enormously productive. This is considered by many the very best table melon for family use. Pkt., 5c.; oz., 10c.; ¼ lb., 25c.; lb., 85c.; 5 lbs. and over, 65c. per lb.

July List o' Things to Do!

* **If you're tempted to use the excess zucchini** from your garden as baseball bats, or have tomato fights, stop! Take that produce to your local food kitchen or charitable food pantry. Why waste it when someone can taste it? (And they'll thank you for it, too.)

Fresh produce is always in demand, and it's always sweet and good to one who's fallen on hard times. (With as many layoffs we've seen, and as tight as the job market is, you probably even know someone who could use that extra help.)

If you're not sure where your local food kitchen is, ask a pastor or priest. Also, many churches hold food drives that you can participate in.

* **Vine borers are a problem** on squash vines. These are caterpillars that drill into the vine several inches above the ground, killing the plant when it blossoms. If you notice a sawdust-like substance on the stem, cut into the injured place and dig out the borer. Then pile up soil above the wound. If you've caught the borer in time, the squash will send out new roots and survive.

Next year when you plant your squash, wrap the stem of the vine with panty hose to keep the borers out.

* **Plant some greens for late-summer salads.** Look for varieties that say on the packet that they're good for summer sowing, or that they are heat tolerant. These include arugula, basil, mustard, bok choi, and gai lan, and some lettuces. Cilantro and spinach is prone to bolting (bursting into flower) in the summer heat, so look for bolt-resistant varieties with these.

Plant the greens under a shade cloth, and sow new plants every two weeks. Keep 'em watered and harvest them promptly.

* **Now that your strawberry patch** has stopped producing, mow it with the mower at its highest setting. This kills off the old plants that are done producing and makes space for the new plants to grow and be ready for next year. Weed the patch, put down some compost and fertilizer, and give it a light mulch of straw. Water the strawberries well to help the plants through the summer heat.

* **Potatoes are setting seeds** – actually, sprouting green acorns on strings is more like it. Pick the seeds off so the potato can direct all its energy to its growing tubers. (You can dig out a few tubers to cook as new potatoes.)

* **Leave those suckers growing on your corn** plants, and don't remove them. Suckers are just a sign that your corn plants have experienced favorable growing conditions. Removing them makes no difference in your yield or in the size of the ears – so let them grow! Sometimes, if the sucker comes up early enough, you might even get an extra ear or two from them.

* **If you fall behind on your weeding,** don't give up the garden. That would be like giving up your living room because it got too

messy and you got tired of cleaning it. Then again, that happens to a lot of people, too!

When trying to catch up on your weeding, the most important thing to do is to make a sizable dent in the garden. If you have a vegetable garden, clear the rows with a weed-eater, then slap down newspaper and hay everywhere. If you have a flower garden, start pulling or hoeing like crazy, even if it means losing a couple of plants. Take the lawnmower to it if you must.

Give yourself fifteen minutes to do the job. That gets you into the garden to start the task – which is often all you need to get motivated to finish the job.

Go after big weeds. Don't bother with tiny weeds. You have to make a hole in the mess. Once you've made progress, your morale goes up, and you're ready to tackle the rest.

* **Blossom end rot of tomatoes** and peppers might make an appearance if the soil moisture has been uneven.

* **To keep insects from chewing up your squash** and cucumber plants, cover them with lightweight floating row covers. Remove the covers once the plants flower, so the bees can get to them.

* **In late July,** sow collards, kale, sweet corn, and summer squash for fall harvest; sow carrots, beets, turnips, and winter radish (daikons) as well. Also, transplant the first crop of broccoli, cabbage, and cauliflower plants into the garden. Make sure they're husky little plants so they're tough enough to take the heat!

* **Harvest sweet corn** when the silks turn brown; harvest onions and garlic when tops turn brown.

* **Blackberries are ripening!! Go pick 'em!**

* **Harvest potatoes** a few days after the tops die. Plant your fall potatoes before August 15.

AUGUST

Late-Summer Planting for Fall Vegetables

To many die-hard vegetable growers, summer offers a new start. By sowing or planting certain vegetable crops in August, you can get a second harvest before (or even after) frost. Imagine fresh vegetables for your Thanksgiving feast ... or even for Christmas dinner, if you put out the right vegetables and mulch them well.

Fall-grown vegetables taste better because they ripen in the cooler weather. You also have a lot less hassle with insects and diseases.

Perhaps you have some seed left over from this spring, or know someone who does. If you need seeds, ask your friendly local nurseryman, and they should be able to scrounge some up for you.

A few nurseries may have cole crops ready to plant – broccoli, cauliflower, cabbage, etc. Call around before you go shopping, just to be safe. If the nursery doesn't have any, ask them if their suppliers can bring some by. (This goes for local nurseries only – the big box stores won't provide this kind of service.)

There's a short window for sowing and planting this time of year. Sow beans, beets, carrots, kohlrabi, peas, zucchini, and lettuce (leaf and butterhead), from August 1 until about August 15. (All dates in this article apply to Zone 5 – if you're in a different zone, check with your local University Extension service for the planting dates that work best for your particular region and climate). Those dates also go for planting cole crops. Sow mustard, radishes, spinach, and turnips from now to August 30.

Be sure to buy "short-season" plants to be certain you can get them harvested before frost. (Don't forget, though, that you can extend the growing season with a few old blankets!)

Prepare your garden. Clean it up and take out any old plants and weeds. Throw out diseased plants and fruits, to avoid re-introducing pathogens into the garden.

Spread compost or aged manure over the ground. Add a pound of 13-13-13 all-purpose fertilizer per 100 square feet of ground before you cultivate the soil. You might till it in, but shallowly. Deep tilling can cause the soil to dry out (more).
With this heat, it might help to water the ground the night before you cultivate, then water it again a day or two before you plant. Otherwise, you will have to take a chisel to the soil.

Sow seeds twice as deep as you do for spring planting, to keep the seed from drying out during germination. Once the seeds are planted, water daily.

To keep the ground cool while the seeds are germinating, shade it by placing a board over your rows, raising it off the ground with two flowerpots or bricks. Or use shade cloth or even sheer curtains to help keep the plants cool. When the seedlings emerge, remove the board, then mulch the ground with a half-inch of grass clippings or straw. Add more mulch once the seedlings are 1 to 2 inches tall.

Sometimes it can be hard to get the crops established. Water them well, and use nylon net to protect the seedlings from insects and the hot sun. At least harvesting plants in fall will be a relief, compared to harvesting in the heat of the summer sun.

Root crops can be left in the ground after frost. Just mulch them heavily for protection, and dig them up as needed through the winter.

In early August, sow beets, daikon radishes, fennel, carrots, green onions, and cole crops – including broccoli, cauliflower, and cabbage.

Later in the month, sow peas, radishes, spinach, cilantro, and Swiss chard for overwintering.

Fix Weeds with Newspaper Mulch

It seems that one day the garden is looking just fine, and all the weeds are under control, but the next day you turn around and your neat rows of vegetables or flowers are suddenly crowded with big, hulking weeds. And in the heat of summer, after a long day at work, you just don't want to melt into a puddle while trying to pull all those weeds up.

Fear not. Here's a way you can knock those weeds down and keep them down. This method doesn't involve chemicals that might affect your produce or damage your plants. It is as cheap as can be – and you don't even have to pull the weeds up.

Newspapers to the rescue!
What you need is a lot of large newspapers (national papers like the *New York Times* or the *Washington Post* are good examples) as well as some mulching materials like grass clippings, pine bark, chopped-up leaves, straw, etc. If you don't have newspapers, go to your local library and ask for any big newspapers from their recycle bin. They'll load you up.

If the weeds are really tall, walk over them to lay them down against the ground. Then open up a section of the newspaper and lay it on the weeds. Each big rectangle of newspaper should be ten pages thick. Lay newspapers down all over the weeds, overlapping the newspaper edges so that light (and weeds) can't get through. As you do this, throw some mulch down to keep the newspapers from blowing around, especially on windy days.

If you have a lot of plants to mulch between, leave the papers folded. You can also tear the newspapers to slide them around the stems of your plants (and this is actually helpful in keeping cutworms at bay).

When the ground is covered with newspapers, get the rest of the mulch and throw it on top. Put a nice, thick layer of mulch down – about three inches – so the next thunderstorm won't pull the pages up.

And then you're done.

This job will take an hour or three, depending on the size of your garden. The really nice thing about newspaper mulching is that when all the newspapers are down with the mulch on top, the garden looks incredibly tidy and clean, compared to the way it looked just a short time ago!

And what's even better? If you've used your 10-layer-thick pages, and you've overlapped them, then you won't have to worry about weeds for the rest of the year. Occasionally a perennial weed might poke through. When that happens, move some papers out of the way just a little bit, dig out the weed, cover up the space with an extra square of newspaper, then re-mulch.

Now your garden is mulched. The newspaper mulch keeps the weeds down, fertilize the soil, cool the roots of the plants in the summer heat, add organic material to the soil, and save water. Earthworms will be active underneath the mulch, tilling the ground for you and adding worm castings, which is pure gold for plants.

A newspaper mulch is a great thing to do for your garden – and for yourself.

August List o' Things to Do!

* **My grandma and I froze some sweet corn.** She uses a simple recipe to freeze her corn. You cut the kernels off the cob until you fill up a kettle. Then you scoop out eight cups of corn into another kettle (or any container that will hold it). To the eight cups of corn you add 1½ teaspoon of salt, ½ cup sugar, and 1 cup water. Then you mix it together and put it in your containers (freezer bags are fine), label them, and put them in the deep freeze. No cooking necessary!

When you take out a packet later, defrost it, then cook the corn until it's tender. It's just as sweet as when you cut it off the cob.

* **Plant lettuce and fast-growing crops** to replace the crops you've harvested.

* **Order your garlic** for September planting.

* **Weed the strawberry patch** and mulch it with compost.

* **Compost and till under residues** from harvested crops.

* **Sow beans, beets, spinach, and turnips** for the fall garden.

* **Spinach might germinate better** if you put the seeds in the fridge for a week before planting.

* **Cure onions** in a warm, dry place for two weeks before you store them.

* **Once your gourds** have achieved adequate fruit set, snip off the growing tips of the vines. This makes the plant direct its energy toward ripening the fruits, instead of vine production.

* **Set out your broccoli,** cabbage, and cauliflower transplants in the garden now.

* **Keep adding to that ever-growing list** of zucchini and tomato recipes.

* **Pick herbs** to use fresh or to dry them. If you keep picking them they'll keep producing.

* **Clean out the cold frame** to get it ready for fall use. It'll be here before you know it.

* **For goodness' sake stay cool!**

SEPTEMBER

Fall is my favorite time of year. Apples are ripening this month, the days are cooling down, and soon we'll move out of summer heat into the wet, gray mornings when the heater starts to kick on. The ragweed will finish blooming and my allergies will settle down, somewhat. But I love how the green trees turn all different shades of green as they get ready for the October color extravaganza. I love the scent of leafmold and soil in the autumn forests. And it's so much easier to get out in the garden now that you can go outside and harvest and prune and pull weeds without melting into a puddle!

Getting Tomatoes Ready for Harvest

Gardeners with tomato vines start getting concerned this time of year. They look at all the green tomatoes that are taking their sweet time to ripen, then look at the calendar and see that the first frosts are around the corner.

Help the tomato plant direct its resources toward ripening its fruit. First, tomatoes need about 40 to 50 days after the blossoms set to ripen. We don't have 40 to 50 days anymore, so take the blossoms and small fruit off the vines. Also, take off any fruit that's deformed or looks funny. In short, choose the best and the most mature tomatoes of the crop, and take off anything else.

Then, prune back any vigorous shoots. Don't take off anything that has a lot of leaves on it, since the plant needs the energy produced in those leaves. Also, try tip pruning. This is where you pinch off the tips of your squash and tomato plants to stop them from growing. The plants then can turn their remaining time before frost toward fruit development.

Immature green tomatoes won't ripen off the vine, but mature green tomatoes will. These are just about as large as a red tomato, and are turning from green to white. Store them in open cardboard boxes and give them about two weeks to ripen at 70 degrees.

The truth about green tomatoes.

Last year I said that green tomatoes don't ripen. I was wrong! Eudora Weldon of Graham, Mo., said, "I want to tell you green tomatoes will ripen. Wrap them in newspapers and store them in a cool place in a cardboard box. Check them in a few weeks. We have tomatoes until Christmas."

So you can do that. If you want to use some of your green tomatoes, make relish out of them, or piccalilli, or fry them up, just like in the book. I think I even remember reading, in one of the Little House books by Laura Ingalls Wilder, about Laura's mother using green tomatoes to make a mock-apple pie.

Mm, pie....

Vegetables in Winter

Imagine growing fresh lettuce, baby spinach and all the fixings for a tender salad while everything else is buried under snow and ice. Imagine extending the growing season by months, even in winter. You can, with a cold frame.

Skeptics might say, "It's too cold to grow plants over the winter." However, plants like spinach, corn salad (or mache), radishes, or lettuce really don't need a lot of heat to grow. Even in the depths of winter, when you push away the snow, you still find a bit of determined green in the grass beneath it. The same is true of these crops. Germination of these plants in a cold frame should be fairly quick, and if you hurry and sow the seeds now, they'll be growing in about a month.

You don't have to worry about freezing temperatures. The crops can freeze with no ill effects. Some of them, like lettuce and mache, can even be harvested while they're frozen, just as long as they thaw out naturally when you bring them in.

My own cold frame experiences began when I was reading Eliot Coleman's *Four-Season Harvest* (Chelsea Green, 1992). Coleman grows crops all winter long by using cold frames as well as unheated walk-in tunnels – a series of curved PVC pipes with greenhouse plastic fastened over the top – that are large enough to stand up in. The plants are grown in the ground inside these tunnels, with a row cover placed on top of them. This double protection keeps the ground temperatures from fluctuating. The crops are still able to get enough sun to grow. Adding more protection would bring the temperature up around the plants, but then the plants wouldn't get the light they need to grow.

Coleman is able to grow crops all winter, and he does this in Maine, though his part of Maine is in Zone 5. However, he has also grown crops through the winter while living in Zone 3 – think northern Montana, North Dakota, Minnesota. If he can do it, so can we.

A cold frame is easy to construct. The frame is a wooden box with no bottom, and it's set directly on the ground. The top of the frame is glass. You can use an old door with glass panels in it, or cast-off storm windows, or a simple sheet of Plexiglas – anything that will let lots of light in. The top of the box should be angled toward the sun to capture sunlight and shed water. You also will need a stick or venting mechanism to prop the box open on warm days.

I built my cold frame with four two-inch pine boards, sizing them so I could fit an old storm window on top when they were assembled. The backboard was 12 inches tall, the front board was 8 inches tall, and the sideboards sloped from the back to the front. The nice people at the lumberyard cut the boards for me. I used nails about six inches long to hammer it together. (Use ear protection, because that gets loud.) Be sure the boards are untreated – some lumber is treated with arsenic, and that's about the last thing you need to be eating with your new crops.

When I put my cold frame together, I quickly discovered that, due to my superior carpentry skills, the storm window didn't quite fit, leaving a half-inch gap that let heat out. I got some plastic grocery sacks and stuffed them into the space, and then I was set. The cold frame worked fine, even with the cheap fix.

Fast-growing crops, such as salad greens, work best in the cold frame, though some root vegetables, such as turnips, can be grown there successfully. Coleman reported that the winter carrots were really sugary – so much so that the kids, when they came home from school, ran out to the cold frames and ate them, calling them candy carrots. The freezing temperatures had turned much of the starch into sugar.

Some crops you might try in a cold frame include carrots, parsley, kale, beets, radishes, lettuce, radicchio and endive, baby leaf spinach, kohlrabi, wild arugula, miner's lettuce or claytonia, buckshorn plantain, and Oriental mustards. In my own cold frame, I wasn't able to get the spinach and mache to germinate (I probably didn't plant the seeds deep enough), but the lettuce came up like gangbusters, and the carrots and beets did just fine.

If you have a taller frame that can accommodate them, you can grow Swiss chard, pak choi, broccoli, peas, and Chinese cabbage.

You can transplant some of your herbs out of the garden into the cold frame, too. Chervil, sorrel, and parsley will do well in the cold frame – or even outside of it, with protection. The same goes for spinach. That will die off in January, or when temperatures get too cold (whichever comes first), but then you can sow spinach seed for an early spring harvest.

You can even sow seeds through the winter. When you eat up all the crops in one area, put down a little seed and start a new crop.

I had some trouble with damping-off disease when growing my seedlings. You might sprinkle Capstan, a fungicide, on the soil around the seedlings to put a stop to the disease. I have also had good results with organic choices like sphagnum peat moss, sprinkled lightly over the ground, or even cinnamon. These change the pH of the soil enough to discourage the disease.

Growth will come to a near-halt from late November through February, when there's too little light to grow on. Just brush snow off the top of the frame during those months. It's fun looking at all the green inside the cold frame when there's a foot of snow on the ground.

If you want to hurry things along during this low-light time, you might experiment with supplemental lighting and see what kind of results that brings. If the cold frame is snug and weatherproof (I'm sure many of you are better carpenters than I am), install some outdoor fluorescent gro-lights along the long sides of the frame. You might put in one red bulb (a "warm light" one) to help with germinating plants, and a blue bulb ("cool light") to help plants grow. Be sure that all the electrical paraphernalia you use is rated for outdoor use, please!

Mache, or corn salad, is Coleman's favorite crop. Mache makes a little rosette of leaves about four inches across and is easy to grow. The tiny plants can be harvested whole and used for salad. Mache used to grow as a winter weed in farmer's fields, where it was picked for salads before it became domesticated. Coleman says you can harvest these and lettuce even when they're frozen, though they'll wilt a little once thawed.

Spinach seems to take the cold a little better than lettuce does, but it may depend on which variety of lettuce you plant, too. Beets are the all-purpose food: both young leaves and roots are wonderful to eat. Coleman grew a lot of carrots because when his kids got off the bus in the afternoon, they'd head right to the cold frames to eat carrots, they were so sweet.

Keep a thermometer inside the cold frame, and one outside of it. Open the cold frame a crack when daytime temperatures rise over 40 degrees, and wide when temps hit 60 degrees. Temperatures over 60 could cause the plants to become susceptible to cold nights. Of course, close the cold frame in late afternoon to hold the sun's heat in through the night. If it turns bitterly cold, throw a blanket over the top.

The only trouble I had with the cold frame was due to cats. On warm days, when the frame was propped open, I'd find a cat sleeping on my tiny carrot seedlings, enjoying the warmth, and I'd have to make all kinds of spitting noises to get it to scram. Thereafter when I vented the cold frame, I took the grocery bags out of the half-inch gap in the back, and I didn't open the front so wide. This allowed air to circulate while keeping cats out.

Next fall, I plan to start the crops for the cold frame in September and October for a December harvest. But even starting the plants in November gave me the earliest lettuce on the block. And it was so good!

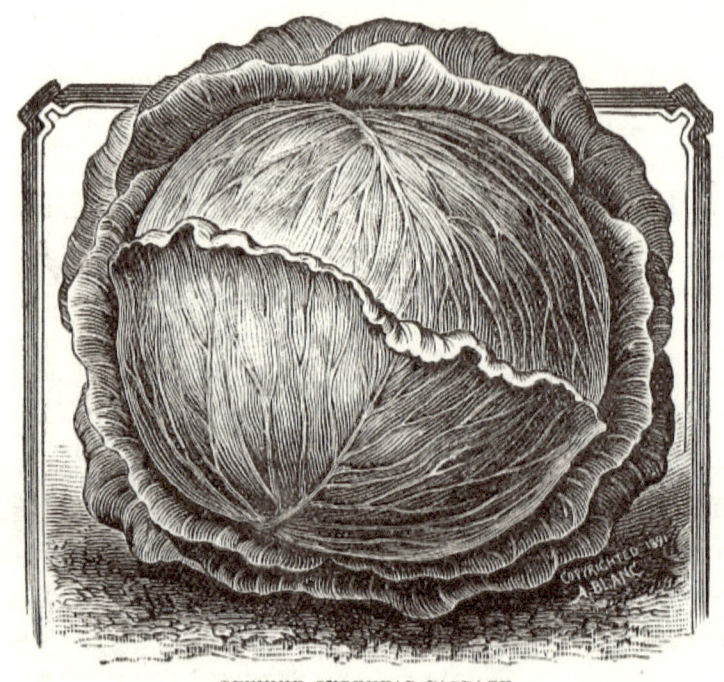
GENUINE SUREHEAD CABBAGE.

Drying Fruits and Vegetables

When I worked at the apple orchard as a college student, I wanted to preserve all the extra apples I brought home, since I was getting them at such a good price. I wasn't crazy about canning them, because I'd seen all the work Mom and Grandma do in high summer as they canned tomatoes, and thought, No thank you! Drying seemed much simpler. So, I peeled and sliced the apples, put them on a cookie sheet, set the oven at a low temp, left the oven door cracked, and a day later I had dried apples.

When I tasted them, I was amazed. Drying concentrated the taste and sweetness of the apple, so each piece was as good as candy to me. I filled a small tin with my apple slices and devoured them all within a week! So much for long-term storage.

Since then I bought a cheap dehydrator that gets a lot of use in summer and fall. I dry tomatoes, bananas, peaches, strawberries, cherries, an occasional squash, as well as quart after quart of apples.

Sun-drying
The traditional method is sun-drying. My great-grandma used to lay sliced apples on a bedsheet on top of one of the outbuildings in the hot sun to dry them. She'd cover them with cheesecloth to keep the insects off, and she'd bring them in at night so they wouldn't reabsorb moisture from the night air.

Some people get a needle and thread and string the sliced pieces up on the thread (the way kids make popcorn strings for the Christmas tree), then space the fruit out on the string and hang it in the sun. Some people use screens to dry their fruit – but don't place fruit directly on a window screen, because the fruit can pick up metal residues from the screen. Lay cheesecloth on the screen first. Small batches can even be sun-dried in the back window of your car -- just be sure everyone in the family is aware of what you're up to so somebody doesn't drive off with your fruit.

Sun drying is most effective when the temperature is close to 100 degrees and when the sun is shining. It's important to bring the fruit in at night so it doesn't reabsorb moisture from the night air

– this makes your fruit moldy. It's also important to keep cheesecloth draped over the food so insects can't get at it, because if a fly gets in there and lays an egg, then you have maggots. You might see some wasps and bees hanging around the area, so watch where you put your hands when you turn over your food as it dries. Sun drying can take from two to three days, depending on what you're drying and the daytime temperatures.

Fruit that has been sun-dried will need to be pasteurized to kill any insects or eggs that may be on the food. There are two methods of doing this. You can put the sun-dried food into freezer bags, then pop them into a deep freeze (one set at 0 degrees or below) for 48 hours. Or you can preheat the oven to 160 degrees, put the food on a tray in a single layer, then put it in the oven for 30 minutes. I prefer the freezer method, simply because I store all my dried food in the deep freeze anyway.

Oven-drying

You can dry small batches of fruit in the oven. Lay the fruit or vegetables on a cookie sheet, keep the oven between 120 and 150 degrees, and leave the oven door open a crack. This method usually takes about a day, and can be done with an electric or gas oven – I dried apples in a gas oven and they did fine. I prefer to use the oven occasionally for late fall fruit, when the days are chilly and I don't want to turn on the heater.

The dehydrator

Naturally, I like the dehydrator the best, since the drying temperature remains constant and I don't have to worry about bugs. My dehydrator doesn't have a thermostat – some do – but

I haven't had anything come out scorched, even when I've left the fruit too long. Also, I love the smell of apples and berries that fill my house when I'm using the dehydrator. I bought a small $20 model with that consists of five trays and an on/off switch, and it's done the job for me for about three years now.

The only drawback to my dehydrator is that I can dry only 11 apples at a time – and when I have a half-bushel of apples (and a full-time job to schedule my drying around), it takes about a week to get them all processed! This year I plan to sun-dry and oven-dry some apples to help get them finished up faster.
Larger dehydrators are available if you really need something that gets the job done!

Preparing produce for drying
In my house, fruits and vegetables like apples, tomatoes, pears, squash, and peaches go through pretty much the same process. I wash them well, throwing out overripe or underripe produce. Larger fruits are sliced ¼ inch thick. Any bruises or bug bites are cut out of the fruit and thrown away. I scoop the seeds and gel out of the tomatoes … or do my best, anyway.

Juicy fruits like peaches and tomatoes are a challenge to slice and lay out. They take longer to dry as well. You can make fruit leathers with these. Mush them up or puree them, then spread them out on plastic wrap to ¼ inch thick. If they are gloppy, add a tiny bit of apple juice to help them pour out. But don't make them too thin or then you'll have to lick them off the plastic, which doesn't work so well. You can oven-dry or sun-dry them. If you use the oven, don't let the heat get over 130 degrees, or you might scorch the fruit leathers or even melt the plastic!

When the top part is dry, peel it off the plastic and flip it over and dry the other side. Then roll or stack and keep them in a plastic bag when they're dried.

Fruits that you can't peel, such as cherries and grapes, offer their own challenges. You can cut cherries in half and dry them that way – you have to pit them anyway – though they get sticky unless you dry them completely. Dusting them will help them keep from sticking. Gen McManiman, quoted in Carla Emery's book, recommends dusting bananas with powdered oats, tomato slices with garlic powder, pears with nutmeg, apples with cinnamon, and so forth.

If you want to dry whole berries, such as cherries, blueberries, grapes, or other fruits with waxy coatings, you will need to crack their skins or else the berries will not dry very well. To crack their skins, get some water up to a rolling boil. Dip the fruit into the boiling water for 30 to 60 seconds, then dip them in cold water. Dry them on absorbent towels, then dehydrate. This boiling-water treatment will get rid of the berries' waxy coating and allows moisture to escape.

Vegetables also work well in the dehydrator: you can dry corn (scraped off the cob), beans, green beans, peas, tomatoes, squash, etc. Vegetables should be sliced thin, because they are low in acid and spoil more easily, so they need to dry faster. In her book The Encyclopedia of Country Living, Carla Emery said that carrots, zucchini, pumpkin, sweet potatoes, squash, and turnips make good snack chips when they're dried in thin slices. You can eat them plain or with a dip.

Pretreatment
University extension services around the country all recommend pretreating the produce you plan to dry. Pretreatment helps to keep the fruit from browning, helps foods with tough skins (grapes and berries, for instance) dry faster, keep vitamins from being lost, and destroys potentially harmful bacteria. For the record, I've never pretreated my fruit before I dry it, since the drying process for me is pretty speedy and I always put my dried fruit in the deep freeze when it's finished and haven't had any trouble yet. However, for people using a slower drying process, or for folks who want to keep vitamins from being lost during the drying process, here are several pretreatments to use.

Ascorbic acid (vitamin C): Pure crystals of ascorbic acid are usually available at supermarkets or drug stores. Stir 2 ½ tablespoons of crystals into a quart of cold water to treat 10 quarts of cut fruit. For smaller batches, use 3 ¾ teaspoons in 2 cups of cold water. Cut the fruit directly into the solution, let it soak for 10 minutes, then remove with a slotted spoon, drain well, then dehydrate.
Citric acid/lemon juice: Use 1 teaspoon citric acid in one quart of cold water, or mix equal parts of lemon juice and water (one cup lemon juice to one cup water). Use this in the same way you use the ascorbic acid solution.

Storage and Usage
Dry the food until it's leathery and dry with no moisture in the middle of the food. After you've dried several batches, you will start to be able to feel whether the slices are dry enough or not. Often, to make the drying process go faster, I'll open up the

dehydrator and take out any fruit that's already dry, opening up gaps in the trays to allow more air circulation.

I dehydrate the fruit until it's quite dry, with no hint of moisture on it. Generally apples are flexible when they're at this point. My peaches, tomatoes, and corn turn out crisp. My bananas turn out very crisp. I have to warn you that when I chew my dried bananas, they stick to my teeth, so basically I just put a banana slice in my mouth and suck on it until it dissolves like a piece of hard candy. This helps keep my munchies at bay when I'm at work and waiting for lunchtime to hurry up.

When the fruit is done, I let it cool off. Then dump the pieces into freezer bags marked with the variety and the date, then let them sit on the counter with the mouth of the bag open just to make sure everything's dry. If I see the tiniest bit of condensation, back to the tray they go. If they stay dry, after a day or two I'll seal the bag and toss it in the freezer. (You can keep the bags in a cool, dry, dark place and they'll be fine, but I like the freezer best.)

You can also store the dried fruit in sealed mason jars as long as you store them in a dark place. Light is bad for dried produce. Don't store dried food in anything metal – but you can put the food in a plastic bag so the dried produce is not touching the metal.

If you want to reconstitute the dried fruit for cooking, soak it for about 12 hours. One pound of dried apples equals 3 ½ to 4 pounds of fresh. As far as dried vegetables go, I love to use them in soup, and I dry them only for this purpose. All I do is throw them into the pot with the rest of my ingredients, then simmer them nicely for 30 minutes, and they come out tasting just fine.

For more information: Your local University Extension service is a wellspring of good information, and they will give you bulletin after bulletin about drying food, and indeed anything else, if you just ask politely. I also love *The Encyclopedia of Country Living* by Carla Emery, which offers the lowdown on drying as well as just about any other aspect of good old-fashioned food production, from vegetable planting to canning to setting up a poultry flock to milk handling and cheese recipes. Lots of good stuff.

September List o' Things to Do!

Here are a few tasks to take care of now that the weather's finally cooling down, to prepare the garden for the fall months before winter comes in. (It feels strange to think about winter right now, though.)

* **If summer weeds have taken over** your garden, it's easy to give in to despair. If only somebody would drive a brush hog through this mess! we cry.

The toughest part is getting started because the mess looks scary and impossible. But once you've started, often you find you're able to make enough progress to feel like you can finish. Little by little, bit by bit is the key.

The most important thing to do is to make a dent in the garden. If you have a vegetable garden, then clear the rows with a weed-eater, or run the lawnmower straight through it. Spray messy garden edges with non-selective weed killer. If you have a flower garden, then grit your teeth and start pulling or hoeing like crazy, even if it means losing a couple of plants.

Give yourself fifteen minutes to do the job. If you feel like working longer, then do so. In fact, you probably will.

Also, go after big weeds only. Don't bother with tiny weeds. You have to make a hole in the mess. Once you've made progress, your morale goes up, and you feel capable of tackling the whole thing.

* **You can still plant** in your garden. Put in mums, pansies, and flowering kale for a little late-season color. Some of the cold-season plants can stay in the ground. Snapdragons and alyssum will keep on giving you color, and will stay green through November, though they'll stop blooming. In the garden, you'll still have parsley, spinach, lettuce, and root vegetables, so don't pull everything up! Unless you really want to.

And of course there's still plenty of time to plant perennials, shrubs, and trees. Their roots will be in the ground, safe from winter's cold, and the roots will grow over the winter, though only a little bit. But when spring hits, these plants will pick right up and start growing, and you won't have to bother with all that spring mud and slush.

* **Tidy up your vegetable garden** if it's gotten wild. Do you really need 15 tomato plants this late in the year? Pull up ten of them. Are the goldfinches finished with your sunflowers? Is your corn all brown? Pull 'em up! And set out a feeder for the finches so you don't break a little bird's heart.

* **Egyptian (top-setting) onion** can be divided and replanted now.

* **This is also a good time to clean up** all those weeds you couldn't get to this summer. Bring along a pair of pruners for the small trees that have gotten a roothold in your gardens. Sometimes, if the ground's really wet after a good rain, and if you're really tough, you can actually pull the trees out, root and all. More often, though, you'll have to clip it at ground level.

Better yet, dig it out with a shovel so you won't have to keep cutting it year after year. Most weeds, though, shouldn't require as much horsepower to pull.

I'm a big fan of mulch. Once you get the weeds pulled, lay down ten layers of newspapers, then cover them with a layer of straw, wood chips, or grass clippings. The areas I mulched in my garden several months ago still looks nice. The stuff I didn't mulch is going to be set on fire. Well, actually no, but don't think I'm not tempted.

This is a good time of year to mulch, anyway. Soon, thanks to your trees, you'll have more mulch than you know what to do with. Start piling dead leaves over your vegetable garden. When you get enough leaves, you can turn your lawnmower loose on them and turn them into a fine mulch. That'll be good stuff to till under when the day comes.

* **Sow your radishes, lettuce, spinach**, and other greens in a cold frame so you can stretch out your fall harvest.

* **Keep picking broccoli regularly** to encourage the plant to form side shoots.

* **Pinch out the tops of your Brussels sprouts** plants. This will keep the plant from growing upward, and helps to plump out the developing sprouts.

* **When the cauliflower heads** are the size of golf balls, tie leaves around them to blanch them.

* **We are getting close to frost date.** Pick off any young tomatoes that are too small and green to make it – this allows the plant to focus on ripening the more mature tomatoes.

* **Sow spinach** out in the garden, too. Later in the season, when the frosts hit, it can overwinter under a layer of mulch and give you a nice spring harvest.

OCTOBER

Pumpkin Season!

I used to work at an apple orchard, sorting apples, stocking shelves, and helping customers. I loved going into the coolers where huge crates of apples were stacked, and I'd climb up to the top crate to bring down the bags of apples we needed. But my favorite memory was on a clear fall morning, when blue mist softened the colorful trees on the hills all around, and my boss was joking and unloading bright orange pumpkins and tossing them to me. The colors – the bright oranges, the soft blues, the endless greens and red and oranges on the hills far away – were just perfect.

So when you get a pumpkin (support your local orchard!) for carving or decorating, here are a few tips.

A pumpkin should be left on the vine as long as possible. Pumpkins can be harvested when the fruit stops growing, and when the skin has hardened and is all orange. Check the pumpkin with your fingernail. A ripe pumpkin will have developed a shell that your fingernail can't pierce. Use a knife or garden clippers to carefully cut the pumpkin off, and leave about two inches of stem attached. Try to leave the vine otherwise undisturbed if there are other pumpkins on it.

By the way, when you harvest, the pumpkin should be at the color you want it to be – unlike tomatoes, green pumpkins won't ripen up for you very well.

When you pick up the pumpkin, pick it up by the sides, NOT by the stem. Sometimes, when the stem has been broken off, diseases can get into the pumpkin, and then it turns to mush.

Sometimes the vine dies off before a pumpkin has completely ripened. You can get pumpkins to orange up a little if the weather cooperates. First, leave them on the vine if you can, but clear away any leaves or debris so the sun can shine on it (this will help orange it up). If frost threatens, cover the pumpkin with clear plastic, anchoring it with rocks. If the day's a cool one, leave the plastic on all day. Temperatures in the 40's slowly but surely kill off pumpkins.

If you bring the green pumpkin into your house or onto the patio, be sure it gets plenty of sunlight and air circulation, and keep turning the green toward the light.

After your pumpkin is harvested, wash your pumpkin off with a weak chlorine solution, using one cup of bleach to one gallon of water. Once the pumpkin is dry, store it in a cool, dry place.

Don't store pumpkins on a cement floor because they tend to rot. Set them on a piece of cardboard instead.

When you choose a pumpkin to make a jack o' lantern, choose one that feels light for its size. There won't be so much meat on these, which makes carving easier.

Before carving, remember that a pumpkin will start to wither in two to four days, so plan accordingly. You might be able to revive a withered pumpkin by putting it in cold water for about eight hours. Put a little bit of bleach in the water to help stop decay.

When you're ready to carve, remember to set down plenty of newspapers or plastic sheeting to catch all the pumpkin brains!

Use an ice-cream scoop to take out the pumpkin seeds. From the inside, crape the side that you're going to carve to about an inch thick to make your work easier.

After you carve the pumpkin, apply petroleum jelly to the cuts and to the entire inside of the pumpkin to prevent rotting and to make it last longer. Or store the pumpkins in large, plastic bags in a cool location to slow down water loss. (Be careful that you don't end up with moldy pumpkins, though.)

You might skip carving a face and simply take the top off the pumpkin. Then fill the pumpkin with late mums from the garden. If the pumpkin's large enough, you might be able to fit a small pot inside it. Then you can still plant the mums.

To bake the seeds, set them out to dry, then cook at 250 degrees for an hour with salt and a little melted butter, stirring them so they don't burn. Also, you can stir-fry them. Seeds are high in valuable vitamins and nutrients, not to mention that eating pumpkin seeds is a fun part of the whole pumpkin tradition.

Cooking the pumpkin is pretty straightforward. Cut the pumpkin in half, scoop out the seeds, put both halves in a glass dish with a little water at the bottom, and pop it into the oven for 45 minutes at 350 degrees, or until the pumpkin is fork-tender. I use the cooked pumpkin for muffins and bread.

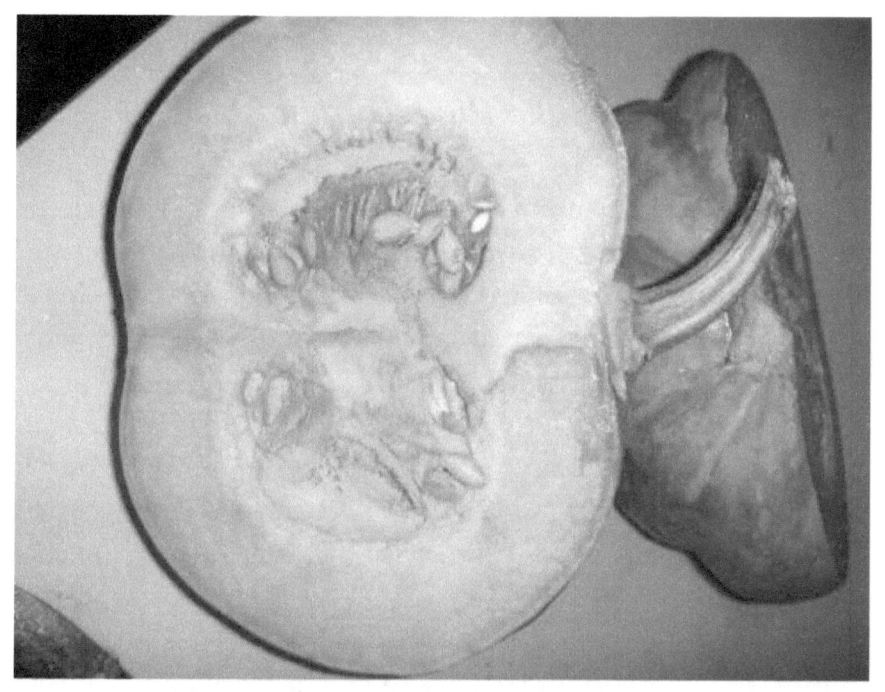

Amendments To Build Rich Garden Soil

I like organic soil amendments. There are all kinds of things that you can put in your soil to make it richer and lighter and fill it with nutrients for your plants. And because the amendments are organic, you can grub around in the soil and not have to worry about weird chemicals on your hands.

Peat moss – This adds no nutrients to the soil, but it's great for making that heavy clay soil lighter. Peat moss is a little pricey, but it's great for growing plants that like acidic soil, such as azaleas or blueberries.

Manure – The important thing here is that the manure must be well-rotted or composted! Let it sit for a year. If you dig it into the soil when it's fairly fresh, the roots of your plant are going to burn. This is because the microorganisms that swarm over the manure to break it down will also break down any roots that are in the area. So now you know.

Leafmold – Boy, I like this stuff. The leaves will take a little while to break down, from one to two years, but when they do break down, you get some nice black soil, as you can see in forests.

In fall, make a little (or huge) bin with chicken wire and four posts and dump your leaves in. Next fall, take the leafmold out and put it on your garden, then fill the bin with the next mess of leaves. If the leaves aren't wholly broken down, work them into the garden soil shallowly and leave them there through the winter, and they should be ready by spring.

You can mow leaves into smaller pieces, though some leaves, like sycamores, will shoot out of your lawn mower as if they're charmed.

Bone and blood meal – Bone meal provides phosphorus, which causes root growth, while blood meal provides nitrogen, which is for green foliage. Have a rabbit problem? Blood meal keeps them at bay (so some folks say). Bone meal, however, should be set around the plants' roots, in the ground, to keep dogs from getting at it.
Read and follow label directions with these, as it's a good idea not to breathe these in.

Compost – Even after all this time, I'm pretty lazy about compost heaps, with all that layering and stirring you have to do. I just throw peat pots and dead plants into a heap and let them sit. Here's an interesting idea: I've heard that if the compost heap is not smelling too good, you can throw several marigolds into the mix and the smell will settle down. Of course I haven't tried it; let me know if it works or not.

Wood ashes – These can go into the compost heap, or you can sprinkle them on or under plants to discourage slugs or bugs. Don't use them too heavily in your garden unless you get your soil tested. Too much ash will change the pH of your soil.

Bradfield fertilizer – This is a brand-name store-bought fertilizer used at Powell Gardens and at the Loose Park Rose

Garden in Kansas City. I used it at the Krug Park Rose Garden and in my own garden. It's got alfalfa meal and molasses (really) and natural sulfate of potash and chicken protein, and it smells pleasantly like rabbit food. The alfalfa contains a chemical that acts as a growth stimulator in plants – it's really great to see the roses sit up and grow when I add this to the soil around their feet, and they do the same for your vegetables.

So there you go. Enough of this and your soil will look like chocolate cake. Your plants will thank you for it.

October List o' Things to Do!

* The first frost (in Zone 5) usually hits about October 15- 20. Are you ready?

* **Protect your plants** from frosts by throwing sheets, blankets, or pillowcases over them. (If you're really good with those blankets, you can carry your tomatoes and peppers into early December.)

* **"Season of mists and mellow fruitfulness,"** John Keats wrote of autumn. It's strange to think that soon we'll be experiencing our first light frost. The days have been so hot and dry that it just doesn't seem possible, yet the nights are cool. As the days get shorter and shorter, we'll be feeling that cool more and more. What a relief (until winter comes and it hits -20 degrees).

This is a good time to start writing down what worked and what didn't work in the garden this year. Consider how you're going to set up your vegetable garden next year. Make a map of where everything grew to avoid planting related plants in the same place. This keeps soil-borne diseases to a minimum. Tomatoes, potatoes, eggplants, and peppers should be rotated every year for this reason.

* **Get your garden soil tested** to see how it's doing, and add amendments if necessary. Fall is a good time to get the work done – a lot of farmers get their soils tested in early spring, so beat the rush by getting it tested early.

* **Try a cover crop on your garden** this winter. After all your crops are harvested, put down a crop of winter rye or legumes to add fertility to your soil. If you decide to cut down the cover crop close to winter, then till it into the ground, you have made your cover crop a green manure – which adds fertility to your soil over the winter.

* **Harvest your winter squash** and pumpkins before frost. Be sure to leave an inch or two of stem on your pumpkins so they don't rot.

* **Harvest all the rest of your vegetables** before that frost hits, too, because they're all going to be done after that. (Unless you have a bunch of them under covers or under mulch – in which case, carry on!)

* **If a big freeze is coming,** dig up your sweet potatoes before it hits.

* **When the shells on your gourds harden up,** and the colors change from green to brown – that's when you harvest them.

* **Root crops like parsnips, beets, and carrots** can stay in the ground into the winter. Mulch them well as the cold weather comes on. The starches in young root vegetables will turn to sugar in the cold weather – making them even sweeter.

NOVEMBER

Collecting Seeds For Next Year

The one nice thing about pulling the dead, frostbitten plants out of your garden is that this is the perfect time to gather seeds off them for next year's planting.

Gathering seeds is a good way to save money and choose the plants that have the traits you want. I've heard about one grower who wasn't satisfied with the orange in most pumpkin varieties, so every year he gathered seed only from the orangest pumpkins in his crop. The next year he'd plant them. By selecting out the best of the crop, year after year, eventually he was able to develop a variety of pumpkin with the bright orange color he was looking for. It's a long-term outlook, but it is certainly worth the effort.

With that in mind, select your seeds off healthy plants, with flowers in a color you like. The seeds or the pod should come off easily in your hand, without the help of pruners. Some seeds are best harvested when they shatter – plants such as bachelor's button, nicotine plant, or poppies throw seeds on the ground, where they scatter in all directions like a broken bead necklace. For these plants, cut off the flowers, stems and all, and shake them, upside-down, into a bucket.

Gather seeds on a dry day, and be certain that the seeds themselves are dry. Wet seeds get moldy in the envelope, and are worthless in the spring. Get all the chaff out of the seeds, if you can, because that will mold, too. Label your seed envelopes clearly, and have a cool, dry place to store them until planting time.

Now, before you go reckless, collecting seeds and whatnot, we need to discuss the difference between a species and a hybrid.

Seeds that are gathered from hybrid plants will give you plants different from its parent. I've gathered seeds from rose moss which had large flowers in all different colors, and the seeds yielded plants with dinky white flowers. Geraniums are the same way (except for the seed geraniums). Seeds from hybrids yield a little plant that won't bloom, or has funny orange blossoms.

Then again, all the marigolds I've planted have been hybrids, but they still produce lovely plants and heaps of flowers that really do look like the parent plant. As for petunias, any seed will produce a huge plant that usually has purple or white flowers. Seeds collected from the second-generation petunias will act and look like the parent plants, producing the same purple and white flowers. The rule of thumb? If you're not sure, try it anyway; you can always pull it up later!

For vegetables and fruits, the seed-collecting rules are a little different. Select an over-ripe, but not rotten, vegetable. Scoop out the pulp and the seeds and place them in a container of water. The good seeds will sink to the bottom, while the chaff and the bad seeds will float. Skim away or pour away the bad stuff floating at the top, shake the jar, and repeat the skimming process several times over the next couple of days. Then carefully pour out the water and collect the remaining seeds.

Lay out the seeds to dry on newspapers or paper towels for several days. The larger seeds may take a week or more to dry completely. Then put them in envelopes, label them, and keep the seeds in a cool, dry place. In winter climates, put the seeds in the freezer for a few weeks, then take them out and store them. If they become damp while they're being stored, they'll have to be re-dried.

If you want to collect seeds from hybrid vegetables, or if you're not sure whether your plant is a hybrid or not, just remember that, with these kinds of seeds, you never know what you're going to get. This is another reason I'm so enthusiastic about heirloom vegetables – they've been propagated by seeds for as long as they've been around, so you always know exactly what you're going to get. (Though, as the fellow with the orange pumpkins did, you can always select for specific traits you want through the years – and create your own heirloom. How about that?)

Taking Soil Samples

Taking soil samples is a topic often overlooked, since it doesn't exactly lend itself to colorful graphics or splashy headlines ("Dirt! Test its cation exchange capacity today!"). Still, a soil test every five years is a necessary part of gardening. One doesn't just look at the soil and say, "Whoa, gotta fix that low potassium level." Instead, one might notice a general listlessness in plants, or odd colors in the leaves, the cause of which one can't determine.

That's why taking a soil sample every two or three years is so important. Soil changes when fertilizer, mulch, sulfur, or lime is added, or when minerals leach out of it. Perhaps you want to put in plants that accept only a limited pH, such as acid-loving plants like blueberries, magnolias, or azaleas, or alkaline plants such as clematis, and you don't know what the pH of your soil is.

Fall is a good time to take a soil sample – and you avoid the spring rush. So, to take a sample, the first thing to do is to get a little box from your local university extension center (or several boxes, if you have a large sample area). Carefully read the instructions they give you. Fill in the appropriate information on the side of that little box. Do this before you dig, since it's hard to write on the box once the soil's inside it.

Get a clean trowel or auger and a clean bucket to collect the soil in. Go to the garden or lawn where you plan to collect the sample. Check the soil for moisture – the soil should crumble easily in the hand. Otherwise, wait until the soil dries.

You'll be taking samples from 10 to 15 random spots around the area. Set the shovel deep into the soil -- over 7 inches deep -- and throw out that shovelful. Then, at the back of the hole (where the shovel's back rested), set the shovel in again, cutting a slice that measures from a half-inch to an inch thick and 7 inches deep. Pull off the thin root-layer of grass at the top and place the soil in the pail. Then repeat that step five or six times within the area of your garden, lawn, or farmland. Take one soil sample for every thousand feet of land. (If you have a huge lawn, you may need two boxes.)

In gardens, take the sample from between the rows to avoid bands of fertilizer that may change the results. If you live near a gravel road, don't take samples near it. The gravel on the road is limestone, and its dust will make your soil test come up highly alkaline, even if the soil is actually acidic.

Finally, take different samples from areas with different purposes. That is, take one sample for the garden, one for the lawn, and one for the orchard, since each has a different fertilization regimen.

When you have taken samples from all of the random spots in your garden, mix all the samples together in the bucket until you have crumbs of dirt – no big clods. Then scoop out enough soil to fill the box, close it, and take it to the local soil testing site. You'll pay a small fee for processing.

Results generally come back in two or three weeks, with a description of your soil's needs that you can start using when next year's planting season begins. You should get a listing of what nutrients are in the soil -- and which ones aren't. The soil test report also will show you the amount of organic matter in the soil, which also is a great part of the soil's structure.

You'll soon get a report on your soil's condition, with suggestions at the bottom to help you out. Then you're on your way to correcting any imbalances in your soil, which your plants will definitely thank you for.

Heirloom Vegetables – A Long and Colorful Tradition

Anybody who has enjoyed tasty vegetables in their garden can find even greater taste, variety, and color through heirloom vegetables – some of which have been around for generations, and even farther back to the earliest days of mankind!

Heirloom vegetables grow true to seed – that is, if you plant a seed from an heirloom, you will get the exact same plant – unlike hybrid vegetables. So when you pick an heirloom tomato, you are picking the same kind of tomato that your great-great-great grandma might have picked in her garden.

History is part of the allure. Many of these heirlooms have a long pedigree. Thomas Jefferson, the third President of the United States, sought out many fruits and vegetables from all over the world to grow in his gardens at Monticello, some of which are still grown there today – varieties that are over 200 years old. You can grow the same vegetables that one of our nation's founders enjoyed. Not bad!

And the sheer variety of heirloom seeds is mind-boggling. If you look at the tomatoes in the Bakers Creek Seed Company catalog, you are in for a treat. I had no idea that we have white tomatoes! Or purple carrots, or even yellow ones! Dainty little white eggplants that really look like eggs! Weird stretchy purple eggplant, and varieties in orange and yellow ... potatoes in all hues and sizes ... lettuces with pretty red speckles ... and beans that have been around since early humans lived in the Fertile Crescent.

In December and January, these heirloom seed catalogs are both a delight and a bane – a bane because you want to try every single one of these plants, and eventually you will have to make actual choices and weed down your huge list of seeds to a measly few. The pain is real, people.

Hybrids – hybridized plants – are different than heirlooms. Hybrids are the result of a cross between two plant varieties. Let's say a grower has two tomatoes, each with a different trait that he wants to combine into one plant. So, the grower crosses these two tomatoes and gets a plant that has both traits. (Technically, when you create a new variety, the process is more complicated, and involves a long process of growing every seed from the hybridized plant, putting the resulting plants through trials in which the lousy plants and bad-tasting fruit from this cross are discarded while the good ones are tested and grown, until the breeder finds a strong, sturdy plant that also bears good fruits that seems to be worth marketing ... anyway, it takes a long time to hybridize a new variety, and then the subsequent marketing also takes time. End digression.)

Now, if you plant the seeds from this hybridized fruit, they will revert back to one of the parents, or they will have crossed with another plant in your garden, yielding something with misshapen fruit, or blight on a stick, or it has pretty little fruits on it but they taste bland.

Hybrids are nice, but there's one difference: Heirlooms are open-pollinated plants – that is, it doesn't matter how they're pollinated, but the seeds will always grow into the same plant, generation after generation. Hybrids can't.

Another good reason to raise heirlooms is flavor. Many commercial varieties of vegetables grown today are not grown for flavor but in order to bear fruit that ships well. Let me repeat that: The vegetables aren't chosen because they taste good – they're chosen because they can travel well for hundreds of miles! Now, if you're going to just pick a couple of tomatoes and carry them inside, you don't need this requirement. You just want something you can stick in your mouth and start chomping on, and you want to have a big smile on your face when you do. Heirlooms fit the bill.

A lot of heirloom varieties have been lost over the years. For instance, in the early 1900s, this country had nearly 7,000 varieties of apples in commerce. These days, that number stands at less than a thousand. The same goes for our garden fruits and vegetables. The movement toward heirlooms is an attempt to bring those numbers back up – to keep the gene pool for these different plants as wide and diverse as possible – and to bring back some of that history we've lost. And also, because this stuff is just flat-out fun to grow. People meeting some of these colorful heirloom tomatoes for the first time are like kids with a box of crayons, checking out all the cool colors. And when they taste these tomatoes … oh boy, then they are sold! Heirlooms – A cool way to enjoy history.

November List o' Things to Do!

* **Tilling the vegetable garden in fall** will expose insect pests to winter cold and kill them off so you don't have to contend with them in next year's garden.

* **Before you start tilling,** throw a bunch of unfinished compost on the soil's surface first so you can get that tilled in, too.

* **Divide your rhubarb plants,** especially if they're overcrowded or haven't been producing worth a darn.

* **Did you know you can store root crops outside** during the winter? Carrots, radishes, turnips, and Jerusalem artichokes can stay in the ground if you throw a nice, deep layer of leaves or straw over them. Then you can mosey out there any old time and harvest them.

* **Clean up your garden.** Burn old crops or compost them. Corn stalks should be burned, if you can, to kill off any overwintering pests. Get rid of fallen, spoiled, or mummified vegetables, too, because they also overwinter pests and diseases that you really don't need in next year's garden.

* **Cover strawberry, asparagus beds,** and your rhubarb with several inches of straw, leaves, or whatever mulch you prefer.

* **Pot up your parsley and chives** and bring them inside to use over the winter. Tuck your rosemary into a sheltered area on the south side of your house (or someplace warm and protected from winter chills, like a mudroom with lots of light) to keep it going as long as possible.

* **Pile your compost materials all together** in the compost pile, and mix them up – balance out the green materials with the brown, if you can – give it enough water to make it work, and let it cook quietly all winter.

* **You can still ripen green tomatoes** over the winter. Lay mature tomatoes in a cardboard box lined with newspaper, leaving space between each one for a little air circulation. Lay a piece of newspaper over those tomatoes, and make a second and

third layer if necessary. Keep the box in a cool room (50 to 60 degrees) to keep ripening slow, but if you want to ripen them faster, move them to a warmer room (70 to 80 degrees) to redden them up. For quick ripening, add a banana to the box.

* **Plant your garlic cloves,** if you haven't already, before the ground freezes. Plant each clove one or two inches deep, six inches apart, in a sunny location.

DECEMBER

An agronomist shows a cabbage grown with compost – and cabbage grown without it! Photo courtesy of @SOILHaiti.

Composting: Learn the Basics

Composting is the art of helping organic materials decay into rich soil. Humus, the name for this type of soil, is compost that has completely decomposed and is "finished."

Sometimes composting seems much more complicated than it should be. To compost, all you need is air, water, nitrogen, and, of course, stuff to compost. Old vegetables, rinds and peels and cores, grass clippings, eggshells, flowers you've deadheaded, carrot tops, bean plants you've pulled up after the beans were all harvested, coffee grounds, old manure, and tea: these are all good for composting.

Stuff you shouldn't compost includes meat, bones, grease, oil, and dairy products. These smell, and they invite vermin.

Generally, a third of your compost should consist of fresh green materials like clippings, weeds and kitchen wastes. Another third should include dead brown materials like straw or leaves. If it's crispy, then it's probably a brown material. The last third of your materials should be unfinished compost or soil. Those provide microorganisms that will break down materials. Setting your compost pile directly on the ground also provides microorganisms as well as earthworms that will help mix things up.

Sometimes you don't have enough green material – just a ton of brown. In this case, mix a good source of nitrogen in with the browns. Blood meal, cottonseed meal, feathers, hoof meal and horn dust, and even hair (pick some up at the barbershop) are all good nitrogen sources.

Even if you don't have an exact 3:3:3 ratio in your compost heap, it's okay. Nature is forgiving. So is compost. If you put together some greens, some browns, and add a little soil to the mix, and stir occasionally, you'll still get some action in the compost.

The best compost seems to consist of three times as much plant material as manure – which is pretty much how nature herself does it.

To make composting go faster, put the materials down in layers – green, brown, and dirt; green, brown, and dirt; and so forth. Don't worry if you don't have the right proportions of greens and browns; the stuff will break down, but it will take a little longer.

Once your pile is made, your compost will need water. When adding dried brown materials, have the water hose handy, and wet them down as you put them in. The compost materials should be moist, like a wrung-out sponge. If the pile gets too wet, it will start to smell. If it's too dry, then it won't break down as fast. Think of the compost pile as a plant you need to keep watered – but don't overwater!

Next, your compost needs air. Every week, use a pitchfork to turn over the materials or stir them up. Some people bury drainage pipes, the ones that are shot through with holes, in the pile to let air in and out. Compost bins need air holes in the sides to let the air circulate. This is crucial for good, fast compost.

If you don't have the strength to stir, get a piece of rebar -- a metal rod used for reinforcing concrete -- or long pipe with which you can stir the pile. Wrap duct tape around the rebar so you don't burn your hand when stirring and aerating the pile. (The faster compost decomposes, the hotter it gets – and a good pile can get very hot!) Plunge the rebar in, stir it, take it out, and tackle another spot.

Most compost bins actually consist of two bins, side by side. One bin can have the compost that you're busy working on and turning, while the other side is accumulating fresh materials for a new batch. When one side of your compost bin has finished materials, take it out and put it on your garden, and start working with the newer materials on the other side.

The University of California has developed a way to get finished compost in 14 days. You get all your browns and greens and shred them, separately of course. (Shredding your materials before putting them in is the key here.) Then layer them in the bin with a bit of soil or finished compost between each layer, wetting the layers (not too much!) as you do. The materials should be as wet as a squeezed-out sponge. Then get out of the way. You will need to stir the compost as it breaks down, but be careful, because it will be very hot! The compost that you get from the 14-day method will be rough, but it will be compost.

Shredding cannot be overrated. You can even compost small branches if you run them through a chipper first while they're green. Dry materials also compost more quickly when they're shredded.

If you have trouble with the compost pile, you might try using worms to break down your compostables. Earthworm farmers and bait shops sell earthworms that you can purchase and put in your compost heap. (Be sure the compost is cold when you put the worms in, of course!) These will help break down the compost, give you lots of valuable worm castings, and they can go into your garden with the compost to help you dig it in. Some worms won't survive the winter, but they'll leave little egg capsules that will hatch into worms next year.

To be sure compost is well done, put a little in a closed baggie for several days, then open it and take a whiff. Well-done compost has an earthy scent. Then take it to the garden.

Some people call compost black gold. You will too, when you see how your plants grow.

Tool Care Tips

Tools are one of the best things to happen to gardeners. You could go out and do your double-digging with an old stick, but you'd probably rather use a clean, sharp shovel. Now it's time to put your tools down for their long winter's nap.

Use common sense through the year – be sure to clean off your tools at the end of a long, hard day in the garden, and wipe down metal parts with an oily rag to keep them in good condition. In winter, there are a few extra things you can do to get everything in order and in tip-top shape for the upcoming season.

* Sand down rough places on wooden handles, then brush them with linseed oil to keep them from drying out and cracking. On metal parts, clean off rust with fine sandpaper or a stiff wire brush, then oil the tool. Motor oil applied with an old rag works fine. Be sure all bolts attached to the handle are tight, not loose.

* Fill a small bucket with sand and a quart of motor oil. When you come in after a day's work in the garden, all you have to do to clean your tools is run them into the sand-oil mixture. The sand helps keep the tools' edges sharp.

* Is there a hole in your garden hose? You can get hose repair kits at the store. You cut out the holey part of your hose, then use the hose repair kit to splice the cut parts back together. Good as new!

* Store all power equipment out of the weather and keep it clean and dry.

* Sharpen the cutting edge on your shovel, hoe, and other tools with a hand file. Before you file, check the angle of the tool's edge, and follow that angle as you file. Use a whetstone to sharpen the delicate edge of your pruners.

* This would probably be a good time to clean out your tool-storage area.

Once you've done all that, you're ready to go for spring.

December List o' Things to Do!

Just because the garden is buried in snow doesn't mean you get to sit back and relax! Whoops, never mind, I guess it does. But when you happen to be doing something other than sitting back and relaxing, make sure it's one of these things from the list below.

* **Throw down some extra mulch** on your perennial vegetables (asparagus, rhubarb, strawberries) once the ground has frozen.

* **Spread compost and bone meal** on your garden and till it in. If you can find rock phosphate or greensand, use that too. These are both a slow-release form of phosphate, and extremely long-lasting.

* **Winterize your power equipment.** Drain fuel tanks or add a gas-stabilizing additive to them. Clean 'em up so they're ready to go once spring rolls around.

* **Clean and oil your garden hand tools, too.** Sand and apply linseed oil to the handles; oil the metal parts; remove rust. You

can also paint the tools a bright color so you can find them if you drop them among the weeds.

* **Start going through your garden notebooks** while your previous year is still fresh in your mind, and make notes to prepare for the new year's garden. Redesign your vegetable garden. Walk around the yard and think about what you want to do next year, then start thinking of ways to carry this out.

* **If you have wood ashes,** sprinkle them around the garden and yard for added fertilization. Only use a little bit unless you are trying to "sweeten" a very acidic soil – wood ashes raise a soil's pH very effectively. It's better to run some of the ashes through the compost heap. This will help neutralize its alkalinity. You might need to add substances like pine needles, which are acidic, to the compost to balance out the alkaline.

* **Even now, walking on semi-frozen garden soil** can cause compaction. Lay boards through your garden to make a path in order to keep the soil light and loose.

Have a good winter and a great rest of the year, and may your garden always be the garden of your dreams.

BOOK RECOMMENDATIONS

Because you need more books and I need more books and books are awesome pretty much all the time.

The Garden Primer – **Barbara Damrosch**
This is the book to read if you are just starting out in gardening. She covers all the basics, and I mean all of them. A great reference for the more advanced gardener, too.

The New Organic Grower – **Eliot Coleman**
This book is really more for the small farmer, and is more advanced, but he goes into a lot of detail about soil building, as well as a number of different ways a small grower can make her plants and soil work for her.

How to Grow Vegetables and Fruits by the Organic Method – **J.W. Rodale and Staff**

The Complete Book of Composting – **J.W. Rodale and Staff**
These last two books are pretty old, but if you can get a copy of them, they are certainly worth the read. Great primers and references for organic vegetable gardeners. Rodale spearheaded the organic movements in America back in the 1940s through the 1960s, and these books draw on the many experiences of his readers.

BOOK PREVIEWS!

If you enjoyed *Don't Throw in the Trowel*, then you'll enjoy my next book: <u>*If You're a Tomato, I'll Ketchup With You: Tomato Gardening Tips and Tricks.*</u>

The National Gardening Association has found that, among vegetable gardeners, tomatoes are their favorite plant to grow. One in three Americans have a vegetable garden, and 9 out of 10 of those gardens have tomatoes in them.

Welcome to the world of tomato gardening.

There's nothing as sweet and good as a sun-warmed tomato fresh from the garden on a hot summer afternoon. It's no wonder that tomatoes are the most popular vegetable in America (though botanically, tomatoes are a fruit). Cordell's book walks you through the steps in raising tomatoes – through starting tomato seeds, planting (and tricks for planting tomatoes early), and staking and caging tomatoes. Readers learn how to fight off diseases and insect pests, decipher the mysterious letters on a tomato tag, how to harvest tomatoes, and how to dry, can, or freeze tomatoes for next year. With plenty of information for advanced gardeners, ready help for beginning gardeners, lots of expert knowledge, and a smidgeon of wit, If You're a Tomato will guide you in the ways of the vegetable garden with a minimum of fuss and feathers. And also with a minimum of weeding. Nobody likes weeding.

Here's a quick sample from this book (complete with pictures).

Tomato Varieties

The National Gardening Association has found that, among vegetable gardeners, tomatoes are their favorite plant to grow. One in three Americans have a vegetable garden, and 9 out of 10 of those gardens have tomatoes in them.

In January and February, when spring fever is really hitting hard, those heirloom seed catalogs inspire dreams of the perfect vegetable garden and the veritable cornucopia of delicious vegetables that it will pour upon our collective tables. Not literally of course, as that would be a mess. But man, those seed catalog tomatoes. It wouldn't be so bad if the seed catalogs could offer taste samples of everything.

When I flip through the Baker Creek seed catalog, which is far and away my favorite, they offer 13 pages of tomatoes (and one page of tomatillos). And so many heirloom varieties. Who knew there was such a variety of tomatoes? Green tomatoes, orange tomatoes, pink tomatoes. Here are the purple, black, and brown tomatoes, which include Black Krim, Cherokee Purple, Paul Robeson, and True Black Brandywine. They have red tomatoes, naturally, two pages of striped tomatoes, two pages of Brad Gate's multicolored tomatoes, some of which are downright psychedelic in purples, yellows, reds, and browns. Then we have blue tomatoes (actually rich purples) and white tomatoes (a very pale yellow, like Bunnicula had been at work on them), and of course yellow tomatoes. I had a Yellow Pear tomato plant that took over half my garden. I can appreciate that kind of vigor in my plants. You can also get peach tomatoes, which have a light layer of fuzz.

Tomatoes also range in size from gigantic beefsteak tomatoes that can weigh up to a half-pound, to the smallest cherry tomato about the size of a marble. You can grow heavy-yielding hybrids or open-pollinated heirloom varieties in different colors, shades, and sizes. You can choose early varieties that set fruit when it's cool outside, mid-season varieties, and late-maturing varieties that will give you the biggest fruits but take 80 to 90 days to do it. Sometimes you'll need about 120 days to get a decent harvest, but hey, at least you get tomatoes!

Tomatoes are so versatile and so good. You can cook them a million different ways or you can eat them, sun-warmed and delicious, straight off the vine. Some people grab a cherry tomato, a leaf of basil, and a slice of mozzarella cheese, and eat them like that.

Yellow Pear tomatoes, an heirloom variety, and also an indeterminate tomato. I had a Yellow Pear tomato plant once that took over the whole garden. It was great.

Determinate vs. indeterminate

I've known about these two different kinds of tomatoes for decades and yet I still can't keep them straight in my head. It's like the difference between flammable and inflammable, kind of.

Determinate tomatoes produce fruit at the ends of their branches. These will stop growing when they are still pretty short.

Indeterminate tomatoes bear fruit along their stems, which keep growing and growing and growing.

If you want a little short tomato to grow in a pot on your patio, get a determinate one.

If you want a Godzilla tomato to take over the world, get an indeterminate one.

Starting Tomatoes from Seed

If you start tomatoes from seed, it's best to start them inside, whether on a window sill or in a cold frame. At any rate, tomatoes need a soil temperature of at least 60 degrees to germinate, though they prefer warmer temperatures, up to 80 degrees. It's a sure bet that your outdoor soil temperature aren't going to get that high any time soon!

Tomato seeds need to be started six to eight weeks before the last frost date.

You can start Early Girl tomatoes (or any early tomato variety) under lights as early as February, then, when the weather is mild enough, transplant to the garden with a Wall o' Waters to help protect the plants against all the frosts.

Traditionally, in Missouri, zone 5, this date has been May 15. With global warming as it is, that date can be moved back to May 1, and even earlier. (Old gardening wisdom is always helpful, but still needs to be updated from time to time.)

So get out your calendar and count back six or eight weeks from your frost date, and that's your sowing date. (Protip: Keep a gardening calendar and notebook where you write down things like sowing dates, the dates you see frost, etc., and use them next winter when you're planning for the next planting year.)

Next, line up your planting containers. Whether you use old egg cartons, Solo cups, flower pots, have them scrubbed (well, don't scrub the egg cartons) if you're reusing them. (Cleaning up the trays/flower pots will clean up any diseases that might be harbored there – diseases that could affect your young seedlings. Use hot water, soap, and a dash of bleach.)

Be sure that, whatever you use, your planter has plenty of drainage holes! This is not negotiable!

For seeds, it's best to use a light seeding mix that is high in vermiculite, though not necessary. A regular "soilless" potting mix is fine. The seeding mix, which is more expensive, also is lighter and easier for newly-germinated seedlings to poke their little green heads through.

When I ran a greenhouse, I started my seedlings in trays and then transplanted them to six packs and four packs later on when they were big enough. For small windowsill operations, this won't be necessary. You can put some soil in a small Solo cup (with drainage holes poked into the bottom), stick two or three tomato seeds in there, and let 'em grow until they're big enough to transplant directly into the garden.

The tomato seeds will germinate more effectively if you have a heat mat under the cups or trays. This will warm the soil with dependable heat, allowing the seedlings to germinate more quickly and grow out more quickly. Just be sure to get a thermostat with the mats so you can adjust the temperature so you don't end up cooking your seeds.

How to plant tomato seeds

Fill your pots, cups, etc. with potting mix, leaving about a half-inch to an inch at the top. Poke two seeds into the middle, about a quarter-inch deep, and cover them. Two seeds are just insurance, just in case one doesn't sprout. When they get bigger, pinch out the wimpier seedling and let the larger one grow.

Gently pack the soil in around the seeds, as seed-to-soil contact is very important for good germination rates.

Sprinkle water on the soil, and keep the soil moist. Don't let it dry out, and don't let it get soggy all the time. Having the soil dry out will kill the seedlings, and constantly wet soil will rot them.

Now one way to keep the soil from drying out is to cover the trays or pots with plastic wrap. This is an easy way to keep the soil moist.

If you use plastic wrap, don't leave the trays in direct sunlight. One time I had the soil covered this way on one of my flats. I came in from one of my jobs and realized that it had been sitting in the sun all afternoon. I ran over and lifted up the plastic wrap – and steam puffed out. Those seeds were roasted!

Once the seeds germinate, though, be sure to remove the plastic wrap, and also have a small fan to keep the air circulating a little around them. Seedling are susceptible to a disease called "damping off" which is encouraged by poor air circulation.

Damping-off disease

I had a bout of damping-off disease in my greenhouse, and it was a mess. Damping-off is a fungal disease that causes newly-planted seedlings to keel over and melt away. It spreads out in a circle, as most funguses do (consider "fairy rings," which are circles of mushrooms on the forest floor), killing off seedlings as it spreads outward.

I haven't had damping-off before, since I keep stuff more on the dry side in the greenhouse, which the fungus doesn't like. However, we had cloudy, cold weather for a whole week, and my trays of seedlings, watered on a Sunday, would not dry out for the rest of the week. No sun, and I couldn't turn on the fans to pull the air through because it was too cold. Humidity was high. All the conditions were right for the fungus to strike.

Then the disease got into the snapdragons I'd just planted and started knocking them out everywhere. I called everyone I could think of for help. Then I took their advice, and it worked.

The best defense is a good offense. Keep a fan running at all times to keep the air circulating. You should feel the air moving through the whole room, but you don't have to turn it up so high that it blows the mice out from under the floor. Keep the plants spaced apart to let air move between them. The fungus likes high humidity and temperatures about 70 degrees. The fan keeps the humidity and temperature lower.

This will break your heart, but get rid of everything that's been infected by the fungus. Dump out the soil and the plants with them, and take the waste outside so spores won't reinfect the plants. As soon as you see the plant keel over, and you know it's not due to being underwatered, out it goes.

If it's a really valuable tray of seedlings, you might dig out the infected plants, isolate the tray from all the other plants, and try a soil drench of Captan fungicide (follow label directions). However, don't bank on saving the seedlings.

Hardening off the seedlings

Once the seedlings get big and husky, and once the weather warms up outside, it's time to harden off the seedlings so they can get acclimated to the weather outside. Plants do better if they have a little time to adjust to the cold temperatures, and the sun, and the wind.

About a week or two before you plan on planting them outside, start moving them outside for a little while. On the first couple of days, water them, then set them outside in the shade in a protected, warm area for an hour or two, then bring them back inside. Then slowly increase exposure to the sun and wind, leaving them out three hours, or four. Then, when you are close to planting time, leave them outside overnight several times (but only if the temperature is supposed to stay over 50 degrees all night).

While you're hardening them off, keep an eye on them to make sure they're not wilting or drying out.

When they're inside, reduce the amount of water you give them, and don't fertilize them until you plant them in the garden.

Don't worry if you miss a day, and don't stress about "not doing it right." Plants are often tougher than we give them credit for, and often there's no real "right way" or "wrong way."

Sometimes regular life gets in the way, so if you can't put your seedlings out every single day, it's okay, just put them out when you can.

If You're a Tomato is available in paperback and ebook. Grab your copy today!

Sample chapter from _Perennial Classics: Planting and Growing Great Perennial Gardens._

CREATING A FOUR-SEASON PERENNIAL GARDEN

You can design a garden that provides interest through all four seasons – spring, summer, fall, and winter. This is where you can really bring all elements of a plant into play – leaf color, form, shape – not just the blossoms.

In spring, summer, and fall, perennials will give you blossoms. The problem is that these blossoms last for a week, maybe a month at the most, before the plant is done blooming. (However, if you cut the perennials back when they are finished blooming, they might reward you with a second rush of blooms.)

One would do well to mix hot-season and cool-season plants in the garden, to make a longer-lasting show of color, instead of watching all the plants turn black at once when frost hits.

A garden for all seasons ... we'd start with some of the earliest-season bulbs, such as snowdrops and crocuses, coming up along the garden border. Then as the season progresses, other bulbs start blooming. With tulips and daffodils, you have early, mid-season, and late bloomers, so you could always have something blooming from March to May – though keep in mind that you'll also have the early-spring perennials blooming and hostas growing gracefully out of the ground.

Then in fall, you also have the option of using cool-season annuals, fall-blooming perennials, and bulbs such as fall crocuses to extend the season. Then, for winter interest, you could have ornamental grasses and the dried flowers of honesty, or money plant; you could have red-twigged dogwood; you could have a Harry Lauders walking stick (*Corylus avellana* "Contorta") with its corkscrew branches. Or grow a witch hazel tree for its yellow blooms in February, or holly with its red berries.

Variegated leaves add color to perennial gardens

To keep the garden colorful, grow perennials with colorful leaves. There's been a big push over the last decade or more to breed plants with variegated leaves, as well as purple, maroon, or chartreuse leaves. Coral bells (Heuchera) were a cute edging plant in the early 1990s, but since then, breeders have been having a field day with this plant, creating a dazzling array of colors in the leaves alone. In one garden catalog, I can find

Heuchera in lime, white, maroon, flame-red, deep orange, chartreuse, bright red, rich caramel, deep burgundy, amber, yellow and pink, purple, and silver. And this rainbow doesn't even include the tiny bell-like flowers, which bloom in bright red, pink, yellow, and white in early summer (though on some plants, the flowers' colors clash with their own leaves). Heucheras were traditionally a shade plant, but the new varieties seem to take more sun than their predecessors.

For a softer, shimmering look, you can plant silver-leaved plants such as Lambs Ears (*Stachys byzantina*). The Lamb's Ears has soft gray leaves that are shaped like lambs ears and are actually fun to pet. Little kids (and grownups) love to pat the leaves. Their flowers are pink, on soft gray stalks, but some people cut down the flowers to keep the soft gray leaves looking pretty.

Artemesia is another gray plant that comes in many different shapes and sizes. You can get Silver Mound (*Artemesia schmidtiana*), which is a soft gray low-growing plant, also a nice plant for little kids (and grownups!) to pat. Or Artemesia 'Powis Castle,' a gray leaved plant that is much taller.

Sedums, a kind of succulent that thrives even when it's dryer than heck outside, and recent years have also seen an amazing proliferation of these plants in the nursery industry. With the sedum, you get more subtle leaf colors than with the Heuchera, but they are still stunning. Many of the sedums are low-growing plants with thick, water-filled leaves, and so these plants can really take care of themselves. They also come in a variety of sizes, shapes and leaf forms. Some plants seem to bristle, though they are thornless, while other plants have soft, rounded leaves. Sedum leaves can be found in shades of silver, deep red, chartreuse, burgundy, while the blossoms (these by the way are lovely) are in all shades and sizes.

Sedum 'Autumn Joy' is a very hardy perennial that bloom in the fall and are visited by monarchs, skipper butterflies, wasps, and bees, and many other insects. It's a lot of fun to watch the diversity of insects and butterflies that visit this plant.

Ornamental grasses can also add color and off-season interest. So even if you have a gap between blooming times in your garden, you can fill the color gap with foliage.

Add shrubs and trees to extend the garden's color through fall and winter

Perennials with colorful leaves give you many options for creating a garden that looks great all year around, even if the flowers you've carefully chosen turn out to have a lull in flowering for a week or two. Of course, you can mix some annuals and shrubs into your garden if you like to add more color and winter interest.

When building a garden, many folks prefer to plant a mixed border – that is, a garden strip that mixes shrubs, perennials, and annuals. This is a solid idea, because shrubs provide four-season interest (in winter, they might have red berries, interesting bark, or red-colored twigs), or can be grown as an eye-catching specimen that the border is structured around. Annuals provide a big splash of constant color from spring until frost kills them off, and perennials offer their own colors and interest during their blooming times. These different kinds of plants providing interest in the border in their own special ways – and it shows the value of diversity.

For instance, a small witch hazel tree will give you blooms in February. Other trees, such as paperbark maple (*Acer griseum*) have papery bark with grays and orangeish reds that add winter interest. Red Twig Dogwoods (*Cornus alba* 'Sibirica'), a shrub with brilliant red twigs, adds interest all winter long, and in late spring has little yellow flowers followed by small blue berries.

For fall, you have fall flowers but also fall color, if you mix in shrubs like burning bush, which turns bright red in the fall, or fothergilla, which turns a lovely shade of yellow. You can also plant stuff with berries on it, or shrubs with red twigs or neat bark for winter interest, and grasses for architectural possibilities. Sea holly (*Eryngium*), a lovely but prickly perennial, has an amazing shape, very formal and Roman, for the garden.

Now, what looks great to one gardener might not work for another. Some gardeners leave ornamental grasses standing in the winter because they like their shape in the landscape. Other gardeners cut them down as soon as they die off in late fall. Some people like to plant hostas all around the trunks of their trees, both for foliage color and to keep the weeds down around the tree. Some people are bored by hostas and prefer only mulch around their trees. At any rate, it's a free country (or it was at press time, anyway).

When you're choosing perennials, also choose plants that offer different kinds of interest besides blooms. Add architecturally interesting plants, those of different shapes, forms, and textures. I always think of sea holly, which is a type of thistle, but it has a stunningly formal shape in the garden, as if it should be on a coat of arms. I have seen these in person, and they are very interesting to look at (though, being a thistle, it is also very prickly). I always want to pair it with pastels – but it would be a good idea to wear rose gauntlets when working with these.

So, also consider mixing up different forms and seeing what this does to the overall effect.

Get your ebook of Perennial Classics today!

Sample chapter from <u>Design of the Times: How to Plan Glorious Landscapes and Gardens</u>

Rules of (Green) Thumb for Garden Design

Once you have all your garden measured out, sit down with your graph paper, let one square equal one square foot (adjust this if you have the graph paper with the tiny squares), and start putting this all down on paper. Have the top of the paper be north, and draw a little arrow pointing that way to look more official.

As you do, you will find that you'll need to keep running out to measure more things in the yard in order to make everything line up correctly. You've measured the house and driveway, for example, but wait, how many feet do you have between the driveway and the east corner of the house? So you measure that. Where exactly does the oak tree sit in relation to the fence and the house? So you measure that. The fence and the house are not lining up correctly. So you remeasure that and try to figure out if you transposed a measurement. The oak tree seems to have inadvertently shrunk, though you'd measured it twice. Maybe it's time to take up drinking. Well, okay, but only in moderation.

Of course, that's the way the old timers did it. (Note: I am not old.) These days, you can fix up a nice landscape plan on your phone using an app, or get a more elaborate program for your laptop that will do more than just move a tree symbol around until it looks like it's placed right.

Then it's time for the big step: drawing a plan. Measure your garden. To keep your plan simple, let one-half inch equal one foot. Draw the outline of the garden on your paper.

Protip: Once you have this step finished to your satisfaction, take this paper to the copier and make several copies, and use these as the rough drafts of your garden design.

Now play with that outline. Consider the height and width of these plants. Keep short plants in front and tall plants in back, and use those pictures you've clipped (whether out of a magazine or found on the internet) to make sure the colors match. Do you want soft colors, such as purple catmint, pink petunias, and silver Artemisia? Or do you want a fiesta of red salvia and "Yellow Boy" marigolds?

Also, consider when your perennials bloom. You may love purple asters and pink sea thrift, but that color pairing won't be happening, because the asters bloom in fall and the sea thrift blooms in spring.

Get the whole family involved with planning your garden!

It's a good idea to put the tall plants in back and short plants in front. Green side up. Match the plants to the amount of sun that's available. Set your shade plants near the trees, while the full sun plants will need to be right out in the sun.

As you draw your plan, generally a good rule of thumb is to arrange them by height – tallest plants in back, shortest ones in the back. Or, as with an island bed, tallest plants in the middle, going to the shortest on the edges. But you can also blend several different varieties of plants that are the same size, the same way as you would blend several different flowers in a flower arrangement, for a good blend of colors and shapes. And you don't have to be exact on regimenting sizes. A garden isn't a lineup of soldiers on dress parade, after all.

You can arrange the plants in any way. You can arrange them in a parterre, a formal setting with neat rows, tidy edged shrubs. Or you can have a wild, natural garden with plants arranged as if they were growing wild. Chances are you will be someplace between these two extremes in your own garden.

Also, to make more of an impact, plant your perennials in drifts of 3, 5, or 7. These groups provide more of an impact than just planting one of every plant (unless you have a specimen plant that's as big as an elephant).

Protip: It's a good idea to have a little out-of-the-way place in your garden where you keep extra plants – those you've picked up when on sale but can't find a place for, plants you've picked up out of curiosity, plants you've gotten from friends and neighbors that don't quite fit into your gardening plan, or things you need to find a proper place for but haven't gotten to yet.

This little garden can get helpful, though. If you have a plant in your regular garden that suddenly croaks, you can grab a full-grown specimen from your little side garden and pop it into your regular garden, if you're so inclined, thereby filling the gap.

You can also keep your cutting garden here, so you can just pop out the back door and cut a few flowers for bouquets inside the house. Then you won't have to swipe flowers from your front gardens and leave holes in it.

Get *Design of the Times* in ebook!

THE END

Me and my worm friends, June 2010

ABOUT THE AUTHOR

I've worked in most all aspects of horticulture – garden centers, wholesale greenhouses, as a landscape designer, and finally as city horticulturist, where I took care of 20+ gardens around the city. I live in northwest Missouri with my husband and kids, the best little family that ever walked the earth. In 2012, when I was hugely pregnant, I graduated from Hamline University with a master's of writing for children; three weeks later, I had a son. It was quite a time.

My first book, **Courageous Women of the Civil War: Soldiers, Spies, Medics, and More** was published by Chicago Review Press in August 2016. This is a series of profiles of women who fought as soldiers or cared for the wounded during the Civil War.

I've been sending novels out to publishers and agents since 1995, and have racked up I don't know how many hundreds of rejections. I kept getting very close – but not close enough. Agents kept saying, "You're a very good writer, you have an excellent grasp of craft, but I just don't feel that 'spark'...." Even after *Courageous Women* was published, they still weren't interested in my books.

In September 2016, I rage-quit traditional publishing and started self-publishing, because I wanted to get my books out to people who *would* feel that 'spark.' In my first year, I published 15 books. This year I plan to repeat that. (When you've been writing novels for over 20 years, you're going to have a bit of a backlog.) I am working my way completely through it and having a complete blast. I love doing cover work and designing the book interiors. I work full-time as a proofreader, so I handle that in my books as well.

And now I'm finding fans of my books who do feel that 'spark.' They're peaches, every one of them.

I'm finally doing what I was put on this earth to do.

There's no better feeling than that.

Thanks so much for reading.

If you like this book, please leave a review on my product page or on Goodreads. Reviews help me get more readers.
Be sure to recommend my books to any of your gardener friends (and even your gardener enemies).

Click here to find out more about all my gardening books, as well as upcoming new releases.

Subscribe to my newsletter
and get a free gardening book!

melindacordell.com

www.ingramcontent.com/pod-product-compliance
Lightning Source LLC
Chambersburg PA
CBHW021441070526
44577CB00002B/239